Differentiation in Military Human Resource Management

Albert A. Robbert
Brent R. Keltner
Kenneth J. Reynolds
Mark D. Spranca
Beth A. Benjamin

Prepared for the
Office of the Secretary of Defense

National Defense Research Institute

RAND

Approved for public release; distribution unlimited

UB
323
.D54
1997

This book may be
recalled before its
original due date.

DATE DUE			
GAYLORD			PRINTED IN U.S.A.

PREFACE

Under 37 United States Code 1008(b), the President of the United States is required to direct a review, at least every four years, of the principles and concepts of the compensation system for the members of the uniformed services. In the three decades during which this legislation has been in effect, seven quadrennial reviews have been conducted, generally under the supervision of the predecessors to what is now the Office of the Under Secretary of Defense for Personnel and Readiness. The Eighth Quadrennial Review of Military Compensation (QRMC) was chartered by the President on January 27, 1995. Its staff was active from early 1995 until mid-1997.

The Eighth QRMC's presidential charter called for designing a military compensation system suitable for the needs of the Department of Defense (DoD) in the twenty-first century. To accomplish this objective, the Eighth QRMC determined that a comprehensive review of military human resource management (HRM) theory and practice would be required. As a result of that review, it concluded that no one HRM system would meet the varying needs of the diverse elements within DoD, and it recommended a contingency approach based on a process of matching HRM systems to strategic organizational objectives.

An important consideration in a contingency approach to HRM system development is the degree to which HRM system outcomes, particularly rewards, are differentiated with respect to individual or group characteristics and behaviors that support organizational objectives. To effectively tailor their HRM systems, system designers must understand the implications of current forms of differentiation and the pros and cons of altering them. In support of the Eighth QRMC's overall objectives, RAND undertook a body of research to examine the potential benefits of greater (or, in some cases, lesser) differentiation. This report provides the results of that research.

This report was prepared under the sponsorship of the Office of the Assistant Secretary of Defense for Force Management Policy. It was prepared within the Forces and Resources Policy Center of RAND's National Defense Research Institute, a federally funded research and development center sponsored by the Office of the Secretary of Defense, the Joint Staff, and the defense agencies.

CONTENTS

Preface	iii
Figures and Exhibits	vii
Tables	ix
Summary	xi
Acknowledgments	xvii
Acronyms and Abbreviations	xix

Chapter One
INTRODUCTION	1
Background	1
Objectives	1
Approach	2
Phase 1	2
Phase 2	2
Phase 3	3
Scope	3
Assumptions	4
Organization of the Report	4

Chapter Two
EVALUATING THE CURRENT MILITARY HRM SYSTEM	5
Major Components of the HRM System	5
Behaviors	5
Personnel Management Processes	8
Compensation (Extrinsic Rewards)	10
Elements Not Depicted	11
How the Three Components Are Linked in a Military HRM System	12
Linking the Specific Elements Within the Components of the Military HRM System	14
Accession Screening	16
Education/Training Selection	16
Evaluation	16
Assignment Selection	16
Labor Supply/Demand	17

Classification	17
Deployment Selection	17
Awards and Decorations	17
Promotion Selection	17
Retention Screening	18
Assessment of the Current Military HRM Framework	18
Evaluation of the Model	19
Perceptions of Military Members	21
Practices in Other Organizations	22

Chapter Three
IDENTIFYING AND EVALUATING DIFFERENTIATION ALTERNATIVES

IDENTIFYING AND EVALUATING DIFFERENTIATION ALTERNATIVES	25
Identifying the Alternatives	25
Make Greater or More-Precise Use of Differentiated Rewards	27
Permit Greater Differentiation in Labor Markets	30
Eliminate Dysfunctional Differentiation in the Current System	33
Evaluating the Alternatives	35
Organizational Interests	36
Assessing Implications of the Alternatives for Organizational Interests	40
Overall Assessments	48
Readiness for Implementation	49

Chapter Four
IMPLEMENTATION ISSUES

IMPLEMENTATION ISSUES	53
Insights from Private- and Public-Sector Implementations	53
Implementation Strategies—The Need for a Phased Approach	55
Demonstration Projects	57
Conditions for Successful Projects	57
Legislative Considerations	60
Specific Cases	60

Chapter Five
CONCLUSIONS

CONCLUSIONS	65

Appendix

A.	MILITARY BASE VISITS: METHODOLOGY AND PILOT DATA SET	67
B.	LINKING HRM TO BEHAVIOR—THEORY, EVIDENCE, AND IMPLICATIONS	101
C.	ACCOUNTABILITY IN MILITARY ORGANIZATIONS	111

References	121

FIGURES AND EXHIBITS

Figures

S.1.	Alternatives As a Function of Potential Effects on Costs and Need for Statutory or Policy Adjustments	xv
2.1.	A Model Linking Behavior, Personnel Management, and Compensation	13
2.2.	Linkages Between Specific Elements in the Military HRM System	15
3.1.	Relieving an Overburdened Promotion-Selection Process	31
3.2.	Alternatives As a Function of Potential Effects on Costs and Need for Statutory or Policy Adjustments	50
4.1.	Alternative Approaches to Accommodate Legislative Constraints	61
A.1.	Degree of Satisfaction with the Level of Military Pay and Benefits	74
A.2.	Perceived Appropriateness of Level of Pay and Benefits As a Function of Referents	75
C.1.	Patterns of Accountability	113

Exhibits

A.1.	Introductory Remarks for Focus Groups	89
A.2.	Protocol for Focus Groups	90
A.3.	Open-Ended Survey (Completed Before Focus-Group Discussions)	94
A.4.	Close-Ended Survey (Completed After Focus-Group Discussions)	96

TABLES

S.1.	Potential Improvements, Alternatives, and Sources of Research	xiv
2.1.	Current and Emerging HRM Practices Not Found in Military HRM Systems	23
3.1.	Potential Improvements, Alternatives, and Sources of Research	26
A.1.	Combat and Noncombat Communities, by Service	68
A.2.	Mean Desirability and Rewardedness of Selected Behaviors	72
A.3.	Mean Size of Effect on Work of Categories of Rewards and Punishments	73
A.4.	Mean Size and Direction of Effect on Work of Levels of Rewards and Punishments	73
A.5.	Proportion of Military Personnel Who Believe That Potential Compensation Factors Should and Do Affect Military Compensation	75
A.6.	Mean Attitudes of Military Personnel Toward Human Resource Management Techniques	76
A.7.	Mean Desirability of Behaviors, by Community	77
A.8.	Mean Rewardedness of Behaviors, by Community	79
A.9.	Mean Size of Effect on Work of Categories of Rewards and Punishments, by Community	82
A.10.	Mean Size and Direction of Effect on Work of Levels of Rewards and Punishments, by Community	83
A.11.	Proportion of Military Personnel Who Believe That Potential Compensation Factors Should Affect Military Compensation, by Community	85
A.12.	Proportion of Military Personnel Who Believe That Potential Compensation Factors Do Affect Military Compensation, by Community	86
A.13.	Mean Attitudes of Military Personnel Toward Human Resource Management Techniques, by Community	88
B.1.	Compensation Theories	109
B.2.	Organizational Theories	110

SUMMARY

INTRODUCTION

The Eighth Quadrennial Review of Military Compensation (QRMC), conducted from 1995 to mid-1997, was chartered by President Bill Clinton to "look to the future and identify the components of a military compensation system that will attract, retain, and motivate the diverse work force of the 21st century" (Clinton, 1995). As part of this process, the Eighth QRMC recognized that compensation is part of an organization's larger human resource management (HRM) system and that such a system can and should be tailored to help the organization achieve its broad goals and objectives. A *strategically* aligned HRM system has three important linkages. First, organizational strategy should inform decisions about the required characteristics and behaviors of people in the organization. Second, desired characteristics and behaviors should inform strategic choices made in the design of HRM systems. Finally, these design choices should shape specific HRM policies and practices.

RAND's research explores one of these linkages by examining the power of current or alternative elements of the military HRM system to influence the characteristics and behaviors of military service members, first modeling how the current system uses rewards to influence behavior, then examining how the system could be improved through greater or lesser differentiation of rewards or other mechanisms, and, finally, examining issues for testing or implementing selected alternatives.

EVALUATING THE CURRENT MILITARY HRM SYSTEM

In seeking to understand the current military HRM system, we modeled that system and sought input from commanding officers, supervisors, and workers gathered during visits we made to selected military installations. In modeling the system, we followed the Eighth QRMC's convention that HRM has three components—personnel management processes (e.g., promotion selection, retention screening, evaluation), compensation and other extrinsic rewards (e.g., accession incentives, allowances, and special pays), and organizational structure (e.g., relative flatness of hierarchical pyramids, degree of self-management)—each of which can be strategically differentiated to help align workforce behaviors with organizational needs. Our model shows how two of these components—personnel management and compensation—relate to individual or group behaviors. Specifically, we show that personnel management processes serve as mediators between behaviors and compensation.

The model illustrates the pivotal roles of promotion selection and retention screening. Most workforce behaviors influence these processes, and most forms of compensation are affected by them. Because of these links to disparate behaviors, ranging from on-the-job effort and performance to participation in individual education programs, there is a danger that promotion selection may be overused as a mediating mechanism. First, it allows individuals to choose their own trade-offs among the types of behavior that influence promotion selections, weakening the capacity of supervisors, commanders, and managers to condition specific behaviors of their choosing. Second, it makes all relationships between behaviors and consequences less clear. Third, because promotion selections may be separated in time from the behaviors they influence, individuals may discount the value of promotion, weakening its capacity to motivate behavior. Fourth, the influence of duty history on promotions rivals that of documented performance (evaluation). The informal, unregulated, and possibly unobserved evaluation systems at work in selecting individuals for promotion-enhancing duty assignments may differentiate between high and low performers more effectively than do highly inflated formal evaluation systems.

We confirmed the insights from the model in our base visits. Although military members strongly favor using individual performance as a basis for promotion, many perceive that promotion processes too often result in poor performers being selected or good performers being passed over. Beyond the problems of over-reliance on promotion selection and retention screening, the modeling and, particularly, the base visits confirmed that the intrinsic rewards associated with military service—serving one's country, meeting challenges, shouldering responsibility—play a large and important role in shaping the behavior of military personnel.

IDENTIFYING AND EVALUATING DIFFERENTIATION ALTERNATIVES

From our assessment of the current system; the implications of three relevant theories—expectancy theory, social justice theory, and transaction cost economics theory;[1] and a review of HRM practices in other organizations (e.g., self-managed

---

[1] *Expectancy theory* offers a model of how rewards for performance affect behavior. The motivating strength of a reward is related to three factors: an employee's expectancy that his or her efforts will produce a worthwhile organizational outcome; an expectancy that the organization will observe the outcome and associate a reward with it; and the valence, or attractiveness of the reward to the employee.

Social justice theory predicts that employees' perceptions of the fairness of a reward system are related to their motivation to perform. Distributive justice theories suggest that people compare their own reward/contribution ratios to the reward/contribution ratios of others. Procedural justice theories suggest that employee behavior is influenced by the consistency and rationality of procedures used to set rewards.

Transaction cost economics theory predicts that the optimal terms of employment relationships (which range from external, spot-market structures to internal labor markets and relational teaming between employers and employees) are context-specific. An important contextual factor in determining the optimal form is complexity: In less-complex environments, spot-market structures are optimal; in more-complex environments, relational structures are optimal.

teams, gainsharing/goalsharing, and multipolar performance appraisals[2]), we identified three major areas in which improvements might be made and identified nine potential improvements within these areas, to which we related 16 policy alternatives. Table S.1 summarizes these elements.

To assess the alternatives, we examined their potential effects on a range of factors that we reasoned would be important to military organizations: productivity, consistency with practices in the broader private and public sectors, compatibility with existing or future organizational and cultural contexts, fairness, compatibility with available accountability mechanisms, and cost (here, limited to very general cost implications rather than specific dollar values). Although all the alternatives were thought to enhance productivity, our productivity assessments are not uniformly positive. Some alternatives affect productivity in multiple ways and perhaps in opposite directions—for example, individual performance rewards (performance bonuses, merit pay raises), which can increase individual productivity but can reduce collaboration within a group.

All of the alternatives we examined increased compatibility with organizational practices observed elsewhere in U.S. society, suggesting that a military workplace with more-differentiated HRM would look more like the private sector. Greater differentiation of HRM systems would also support greater flexibility in the structures and cultures of the military services or elements within them. The price, in some cases, might be a weakening of command and control and the military's culture of shared sacrifice and service.

In terms of fairness, double standards seem to apply in many cases, with the two standards almost always yielding conflicting assessments. As for accountability, assessments in a few cases indicated that a lack of alternatives to traditional supervisory oversight could present problems. Finally, in costs, some alternatives have no effect, some have increases that can be readily offset, some entail risk premiums (increases in *expected* compensation needed to offset reduced certainty in compensation, as is typically the case in pay-for-performance systems), and some entail major administrative or payroll increases.

As an additional step in assessing the alternatives and in helping to think about implementation issues, we divided the alternatives into the categories shown in Figure S.1. The alternatives can be distinguished by whether they have high or low potential cost implications and whether they require limited or significant policy or statutory adjustments. Those in the bottom-left quadrant are within the military services' current purview and could be implemented now; those in the top-left quadrant appear to require significant budget adjustments, requiring detailed cost/benefit analyses beyond the scope of this research; and those in the bottom-right quadrant warrant that careful and deliberate implementation procedures be established to make them available by the services.

---

[2]*Gainsharing* is an employee bonus system tied to operating efficiencies. *Goalsharing* is a bonus system tied to organizational goal achievement. In *multipolar performance appraisals* (also referred to as "360-degree evaluations"), peer and subordinate inputs are combined with traditional supervisors' evaluations.

Table S.1

Potential Improvements, Alternatives, and Sources of Research

Potential Improvement	Policy Alternatives	Sources of Research
Make Greater or More-Precise Use of Differentiated Rewards		
Increase strength of performance/promotion linkages	• Reduce weight of human capital development in promotion	Expectancy and social justice theories; model analysis
	• Reduce weight of occupational differences in promotion	
	• Increase validity of subjective evaluations through multipolar performance evaluation	
Provide additional performance and productivity incentives	• Establish intragrade merit pay raises	Expectancy theory; practices in other organizations; transaction cost economics theory
	• Establish individual performance bonuses	
	• Establish gainsharing or goalsharing	
	• Introduce self-managed teams	
Increase incentives for human capital development	• Provide competency-based pay for education, fitness, marksmanship, etc.	Expectancy theory; model analysis; practices in other organizations
Further differentiate retention incentives and controls	• Increase special pays/bonuses (to offset less occupational differentiation in promotions)	Social justice theory; model analysis
Permit Greater Differentiation in Labor Markets		
Externalize labor markets	• Relax lateral-entry rules	Transaction cost economics theory; practices in other organizations
	• Increase portability of retirement benefits	
Differentiate labor-market governance structures	• Permit HRM policies and practices to differ across organizations and functional communities	Transaction cost economics theory; practices in other organizations
Eliminate Dysfunctional Differentiation in the Current System		
Eliminate differentiation based on dependency	• Eliminate differences between benefits for members with/without dependents	Social justice and expectancy theories; base visits
Remove mobility disincentives	• Eliminate withholding of the basic allowance for subsistence (BAS) from enlisted members who are deployed	Social justice and expectancy theories; base visits
	• Fully compensate members for permanent change of station (PCS) moves	
Resolve perceived enlisted pay fairness issues	• Raise enlisted pay levels	Social justice theory; base visits

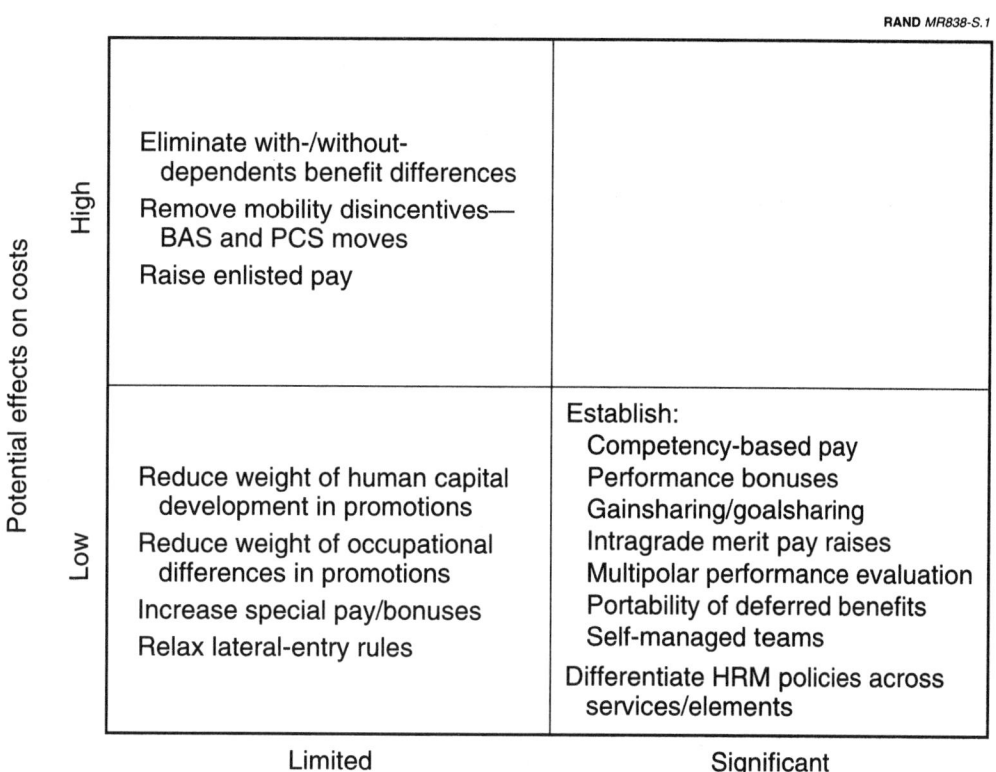

Figure S.1—Alternatives As a Function of Potential Effects on Costs and Need for Statutory or Policy Adjustments

IMPLEMENTATION ISSUES

The alternatives in the bottom-right quadrant would, in most cases, move military HRM systems from less to more differentiation, possibly requiring significant cultural change. Given the experiences of private-sector (and a few public-sector) organizations that have made similar shifts, we propose using a phased approach for implementing these alternatives—moving from conceptualization through pilot testing to more-widespread implementation of new alternatives.

In conducting *demonstration projects* (a term used to describe many federal government pilot tests of new management alternatives), organizational and HRM system managers must maintain a balance between enthusiasm for a new initiative and objectivity in evaluating it. Factors that designers of demonstration projects should consider include the need for promoting understanding and acceptance at all organizational levels, the risks and consequences of failure of a demonstration project, the availability of measures of outcomes and environmental variables, and analytical needs such as control groups or baselining periods.

A special provision of law gives federal agencies the authority to demonstrate civil service HRM policies and practices that would otherwise require a change of law. No similar flexibility is provided for legislation governing military HRM. We propose that the civil service demonstration authority be used as a model for a similar provision for military HRM. We further argue for the advantages of using a "trailblazer" demonstration project—an initial project with great natural momentum—as a vehicle to obtain the needed demonstration authority, and discuss the merits of a likely trailblazer: performance incentives for the acquisition workforce.

CONCLUSIONS

We conclude that military HRM works well in motivating many desired behaviors. Nonetheless, we see that some functional communities today might benefit from greater differentiation and that future environments may demand even more differentiation. We see the potential for differentiation being higher outside of core combat activities, because shifts from hierarchical channeling of authority—a necessary element in many forms of differentiation—are not compatible with other exigencies in a combat environment. We offer several cautions, including a need to avoid undermining a powerful and well-focused system of intrinsic rewards and a need to support increased differentiation with better alternative accountability mechanisms.

ACKNOWLEDGMENTS

We are indebted to Bob Emmerichs, Executive Director of the Eighth QRMC; BrigGen Orest Kohut, Deputy Director; COL Dave Moore, our project monitor; and others from the staff of the Eighth QRMC for their assistance and support during the course of our study. Maj Mark Casey, Maj Oregon Emerson, MAJ Roy Burton, and LCDR Jeff Taub, all of the Eighth QRMC staff, accompanied us on field visits to installations of their respective services and greatly assisted us in interpreting what we learned during those visits. Our RAND reviewers, Glenn Gotz, Bruce Orvis, and Beth Benjamin (who also coauthored one part of the report), provided many thoughtful comments. Our communications analyst, Paul Steinberg, helped us greatly in weaving the parts of the research into a coherent whole. Finally, Marian Branch carefully edited the final version of the text.

ACRONYMS AND ABBREVIATIONS

AFSC	Air Force Specialty Code
ANOVA	analysis of variance
BAQ	basic allowance for quarters
BAS	basic allowance for subsistence
C^4I	command, control, communications, computers, and intelligence
DAWIA	Defense Acquisition Workforce Improvement Act
DoD	Department of Defense
ERISA	Employment Retirement Income Security Act of 1974
FICA	Federal Insurance Contributions Act
GAO	General Accounting Office (U.S.)
HR	human resource
HRM	human resource management
JCS	Joint Chiefs of Staff
JROC	Joint Requirements Oversight Council
MOS	Military Occupational Specialty
MWR	morale, welfare, recreation
NCO	noncommissioned officer
NEC	Navy Enlisted Code
OCS	officer candidate school
OPM	Office of Personnel Management (U.S.)
OSD	Office of the Secretary of Defense
OTS	Officer Training School
PCS	permanent change of station
QOL	quality of life
QRMC	Quadrennial Review of Military Compensation
ROTC	Reserve Officers' Training Corps
SORTS	Status of Resources and Training System
TCE	transaction cost economics
UCMJ	Uniform Code of Military Justice
USC	United States Code
VHA	variable housing allowance
YOS	years of service

Chapter One

INTRODUCTION

BACKGROUND

The Eighth Quadrennial Review of Military Compensation (QRMC), conducted from 1995 to mid-1997, was chartered by President Bill Clinton to "look to the future and identify the components of a military compensation system that will attract, retain, and motivate the diverse work force of the 21st century" (Clinton, 1995). More specifically, it called for designing a military compensation system suitable for the needs of the Department of Defense (DoD) in the twenty-first century. To accomplish this objective, the Eighth QRMC determined that a comprehensive review of the theory and practice of military human resource management (HRM) would be required. During that review, it developed an appreciation for strategic design of HRM systems—the process of aligning HRM policies and practices with organizational goals and objectives.

The Eighth QRMC recognized three important linkages in a strategic HRM design process. First, organizational strategy should inform decisions about the required characteristics and behaviors of people in the organization. Second, desired characteristics and behaviors should inform strategic choices made in the design of HRM systems. Finally, these design choices should shape specific HRM policies and practices.

OBJECTIVES

The purpose of RAND's research was to explore one of these linkages by examining the power of current or alternative elements of the military HRM system to influence the characteristics and behaviors of military service members. Organizations contemplating the QRMC's recommended strategic approach to HRM design would find an understanding of the current capacity of the system to influence behavior to be a useful starting point. Further, insight into how this capacity could be improved would help those organizations identify options for achieving more strategically aligned HRM policies and practices.

The central focus of our research was the capacity of the military HRM system to differentiate rewards with respect to characteristics and behaviors of people in the organization. If HRM systems are to be effective in engendering desired characteristics and behaviors, they must reward more-desired behaviors and punish or discourage less-desired behaviors. This dichotomy implies that greater differentiation of rewards would occur in the HRM system. However, military HRM systems have traditionally featured very circumscribed degrees of differentiation, stressing instead uniformity and consistency. It was important for us to understand the basis for this stress on uniformity and consistency and what might change if the system becomes more differentiated.

APPROACH

Our research occurred in the following three phases.

Phase 1

The first phase was devoted to gaining a better understanding of the important elements in our area of interest, which include desired behaviors, available rewards, and mechanisms that link them together. To gain an understanding of desired behaviors, we first compiled a list of the behaviors that are customarily targeted by military HRM systems (e.g., attraction and retention of the right kinds of people) and supplemented this list with information gained from visiting military bases[1] and innovative firms.[2] We then catalogued the rewards available in current military HRM systems and the HRM processes that mediate the distribution of these rewards. Finally, we constructed a model depicting the linkages between desired behaviors, HRM processes, and available rewards.

Phase 2

In the second phase of our research, we looked for ways to enhance the ability of the HRM system to influence behavior by increasing or sharpening the differentiation of rewards. In this phase, we gained some insights by examining the model constructed in the first phase. We gained other insights by considering several bodies of theory about the relationship of HRM system features to individual behaviors—specifically, expectancy theory, social justice theory, and transaction cost economics theory.[3]

---

[1]As part of our research, RAND researchers visited one installation from each of the military services, and conducted focus groups and surveys among military members. The methodology and survey forms used are presented in Appendix A. Results of this research appear throughout the document in support of specific points.

[2]Specifically, RAND selected six public- and private-sector organizations for in-depth research and analysis. The sample of organizations visited—a state government, a global media/telecommunications company, two defense electronics firms, an automobile producer, and an equipment/document-processing company—had all undergone a major change in organizational structures and work practices in the past few years and were consciously using HRM policies to support and effect their change efforts. Results of this research appear throughout this report in support of specific points.

[3]The three theories, discussed in some detail in Appendix B, may be described briefly as follows:

Expectancy theory offers a model of how rewards for performance affect behavior. The motivating strength of a reward is related to three factors: an employee's expectancy that his or her efforts will produce a worthwhile organizational outcome; an expectancy that the organization will observe the outcome and associate a reward with it; and the valence, or attractiveness of the reward to the employee.

Finally, we observed what other organizations are doing—by surveying what is reported in the literature and by making first-hand visits to innovative firms. From all these efforts, we produced a list of alternatives having the potential to enhance the current system.

After compiling the alternatives, we considered their effects on a variety of factors that we or the Eighth QRMC had identified as being important. It was here, for example, that we explored the outcomes of less uniformity and less consistency in the distribution of rewards. Since empirically testing alternatives was beyond the scope of our research, our work here was somewhat hypothetical, consisting largely of predicting the probable outcomes based on theory, our own observations, and feedback from military members in the field.

Phase 3

Recognizing that our predictions of outcomes were largely hypothetical, we saw the need for an implementation strategy that permitted testing and assessment of some alternatives on a limited scale prior to full implementation. Outlining such an implementation strategy, the third phase of our research, required learning from organizations that have successfully moved to greater differentiation in their HRM systems and applying those lessons in the bureaucratic and legislative environment peculiar to military systems.

SCOPE

Specifically excluded from this research was any detailed examination of the level of compensation of military personnel relative to that of employees in other organizations. Along with other research organizations, RAND has extensively studied this aspect of compensation policy. Readers are referred, for example, to a discussion of civilian wage indexes in Hosek et al. (1992) and other studies examining the effects of compensation levels on recruiting and retention. Also excluded from this research was an extended examination of economic theory to optimize retention and motivation within the existing structure of military compensation, a topic recently explored in, for example, Asch and Warner (1994). We chose, instead, to focus with less depth on a broad inventory of behaviors required in military organizations, because we considered that breadth rather than depth better served the strategic orientation of the Eighth QRMC.

Social justice theory predicts that employees' perceptions of the fairness of a reward system are related to their motivation to perform. Distributive justice theories suggest that people compare their own reward/contribution ratios to the reward/contribution ratios of others. Procedural justice theories suggest that employee behavior is influenced by the consistency and rationality of procedures used to set rewards.

Transaction cost economics theory predicts that the optimal terms of employment relationships (which range from external, spot-market structures to internal labor markets and relational teaming between employers and employees) are context-specific. An important contextual factor in determining the optimal form is complexity: In less-complex environments, spot-market structures are optimal; in more-complex environments, relational structures are optimal.

In addition, in many cases, useful differentiation will require that central human resource managers cede control of processes or outcomes to line managers, resulting in less uniformity of treatment of military members. We recognize that this lack of uniformity might make it more or less difficult for organizations to enforce equal-opportunity and -treatment standards. However, we had no basis for predicting the effects of the alternatives in this area and so did not include this as a specific area of interest in our research.

Finally, as part of our analysis, we assess adjustments needed in a number of factors, including cost, as a result of alternative HRM policies or practices. Developing full cost estimates was beyond the scope of our research. However, for each alternative, we indicate the direction in which costs are expected to go and the expected order of magnitude of those costs.

ASSUMPTIONS

The underlying theoretical foundation carried into this research is that more-rewarded values, traits, and behaviors tend to occur more frequently than do those that are less rewarded. This relationship has been firmly established through behavioral research (Watson, 1919; Skinner, 1938). In this study, we do not examine the basic research supporting this relationship, but we do make extensive use of more-recent theory that applies this behavioral relationship in an HRM context.

In writing this report, we assumed that readers are generally familiar with elements of current military HRM systems. Thus, while we have catalogued the major elements of these systems so that they could be related to our topic, we have described them only in very broad terms.

ORGANIZATION OF THE REPORT

Chapter Two describes and assesses the current military HRM system. Chapter Three presents the alternatives identified from the evaluation in Chapter Two and evaluates those alternatives. Chapter Four provides our recommended approaches for testing or implementing the most-promising alternatives. Chapter Five presents our conclusions.

Appendix A discusses the methodology used in the focus groups and surveys conducted in the pilot study at four military bases, the pilot data set, and copies of the survey forms. Appendix B describes the three theories used in the study and how we selected them from a larger body of theories about the relationships of HRM systems to individual behaviors. Appendix C provides background information on the issue of accountability, which is one of the factors included in assessments of the effects of our alternatives.

Chapter Two

EVALUATING THE CURRENT MILITARY HRM SYSTEM

To determine which alternatives for differentiation in the military HRM system make sense for the military in the future, we began our research by assessing what the current system looks like and how it works.

In this chapter, we start by building a model of the current military HRM system, which entails identifying the specific elements of three major components of the system—behaviors, personnel management processes, and compensation—and explaining the linkages between those elements. Then, we evaluate the completed model. In addition to modeling the current system, we used feedback from the series of focus groups and interviews with commanding officers, supervisors, and workers we conducted at military bases as part of the evaluation.

MAJOR COMPONENTS OF THE HRM SYSTEM

Behaviors

Within this study, we define *behaviors* as the actions of individuals in response to their environments. Behaviors can be described by placing them in broad categories. Within a category, substantively different behaviors can be described. Categorically, the behaviors of interest to an organization might include making labor-supply decisions (e.g., joining, staying, and leaving) or expending effort in a productive capacity. Within a category, behaviors can be substantively differentiated. For example, within the category of labor-supply decisions, *staying* is substantively different from *leaving*. Within the category of effort or performance, a wide range of subcategories as well as substantively different behaviors can be distinguished, at varying levels of detail. For example, individual effort and teamwork might be important subcategories. With respect to teamwork, substantive behaviors, such as cooperativeness and selflessness, can be identified. Finally, substantive behaviors can be measured—groups or individuals may display none, a little, or substantial amounts of a given behavior.

Ideally, HRM systems evoke optimal amounts of substantively desired behaviors in the necessary categories. In this study, we have attempted to identify all the important categories. We have also identified many apparently useful substantive behaviors. But our cataloguing of substantive behaviors was not exhaustive, nor did we develop any insight into the optimal amounts of substantively desired behaviors. Thus, our model of the HRM system identifies behaviors only at a broad categorical level.

As a starting point in categorizing behaviors that are important to military organizations, we reasoned that HRM systems are designed to obtain desired behaviors from workers. Categories of desired behavior can be identified as worker responses to HRM system functions, such as attracting, selecting, classifying, developing, assigning, evaluating, promoting, compensating, motivating, retaining, and severing. We initially developed a list of categories through this deduction; then, we verified and expanded that list through interviews with commanding officers, supervisors, and workers at military installations.

Ultimately, we arrived at the eight categories of behavior discussed below: four labor-supply behaviors (joining, assignment sorting, occupational sorting, and stay/leave decisions), human capital development, effort and performance behaviors, professionalism at work, and responsibility at home.[1]

Labor-Supply Behaviors. Labor-supply behaviors constitute a broad category of responses to attracting, selecting, classifying, assigning, and compensating functions. Initially, potential workers respond to the function of attracting by seeking employment or membership. In a military context, individuals respond to recruiting by *joining* a military service through enlistment or commissioning. Later, workers continually reevaluate their alternative opportunities and take *stay/leave actions.*

Within an organization, more-specific labor-supply behaviors occur in response to occupational and placement opportunities. We label these behaviors *occupational sorting* and *assignment sorting*. *Occupational sorting* refers to developing preferences for, and seeking duty in, specific military specialties. *Assignment sorting* refers to actions taken by individuals to develop preferences for, and seek duty in, specific units, positions, or geographical locations. Assignment sorting can have lateral or vertical vectors. The lateral vector involves choosing among alternative assignment options. The vertical vector involves choosing and seeking a level of responsibility and has a meaning similar to that of *ability sorting*, a term found in a related body of RAND research (Asch and Warner, 1994).

Human Capital Development. *Human capital development* describes behaviors that maintain or increase an individual's current or future productive capacity. Examples include participating in education programs and individual or unit training activities, gaining experience in key functions, and honing physical skills such as fitness or marksmanship.

Effort and Performance Behaviors. *Effort/performance* constitutes another broad category, encompassing the behaviors directly involved in producing organizational outputs. Within this category, we view *effort* as a more quantitative descriptor of behavior: The number of hours worked in a period of time is an observable measure of effort, although effort may also have an intensity that is not easily observed.

---

[1] We also identified a potential ninth category: family formation. Military compensation and housing policies differentiate sharply with respect to dependency status. *Family formation* describes the behavior that changes dependency status and hence changes these entitlements. However, we chose not to include this behavior in our model because it does not seem to be one that military organizations intentionally seek to engender through their HRM systems.

Applicability and effectiveness of the effort expended are dimensions of *performance*, which we see more typically described qualitatively, although metrics can be developed for them. For most of our research, we found it unnecessary to distinguish between quantitative and qualitative aspects of productive behavior. Thus, we generally use the combined term "effort/performance" in this report. When we needed to distinguish quantitative and qualitative aspects, we use the terms separately.[2]

Of the categories of behaviors listed here, effort/performance is perhaps the broadest and most complex. It occurs at individual and group levels, has global and local aspects, and has immediate and long-range dimensions. At an individual level, it includes the behaviors of both leaders (leadership style; exhibiting concern for the welfare of subordinates) and followers (loyalty; acceptance of military discipline and hierarchy). At a group level, it includes elements of teamwork and collaboration. Global aspects—those valued among virtually all military members—include subjecting oneself to the risks of combat and other arduous duties. Local aspects—effort/performance valued in specific organizational, occupational, or functional contexts—range from salesmanship among recruiters to heightened security consciousness among those working with sensitive compartmented information. In an immediate sense, effort/performance encompasses responses to specific challenges in day-to-day situations. In a long-range sense, it includes repeated patterns of behavior, such as tolerance for regular and frequent deployment or regular and frequent family moves, which are valuable to the organization and become part of its culture.

Professionalism at Work and Responsibility at Home. *Professionalism at work* and *responsibility at home* are two categories that capture certain organizational citizenship, ethical, and image-related behaviors that are important to the organization but that are often not directly related to productivity. Neither of these behaviors was among our initial list of behaviors; both were suggested in interviews and surveys with military personnel. *Professionalism at work* includes such traits as integrity, loyalty, and nondiscrimination, as well as more-incidental behaviors, such as punctuality, personal grooming, physical fitness, and proper use of military courtesies. It includes an acceptance of uniquely military responsibilities, including arranging of one's personal affairs in order to maintain a responsible level of availability for contingencies, including shift work, extended hours, field training exercises, and deployments. *Responsibility at home* relates to marital fidelity, financial responsibility, and avoidance of substance abuse.

Behaviors and Organizational Objectives. The behaviors identified above can be viewed as being instrumental in helping the organization to reach larger objectives—specifically, in this context, maximizing productivity. Labor-supply behaviors put productive capacity where it is needed. Human capital formation increases productive capacity. Effort/performance behaviors translate capacity into actual productivity.

---

[2]This distinction was needed in our field research at military installations, as described in Appendix A. In surveys used as part of that work, we described *performance* as "how well you do your job" and *effort* as "how hard you work."

Productivity is defined classically as a ratio of outputs to inputs, of which labor is one. Hatry and Fisk (1992) argue that, in the public sector, the most important final products may not be measurable in physical units of *output* but rather in terms of less-well-defined *outcomes*. Police productivity, for example, can be based on measures such as number of arrests but might be more appropriately based on some indicator of crime suppression.

In a military context, *output* can be measured in terms of sorties flown, tank-miles driven, or ship steaming days, or in terms of the more-detailed operational, logistics, and administrative tasks that underlie these outputs. For many military units, the most important *outcome* is effectiveness in wartime (or in peacemaking, peacekeeping, and similar operations other than war), an outcome that is infrequently observed. A more immediate and ongoing outcome is readiness.

Readiness, as with productivity generally, is an outcome influenced by multiple input factors, including personnel. Unfortunately, outcome measures of readiness are apparently beyond the current state of the art. Instead, readiness is typically evaluated by measuring availability of the inputs required to generate it (Moore et al., 1991). For example, in the Joint Chiefs of Staff (JCS) Status of Resources and Training System (SORTS), a personnel component of readiness is expressed as the percentage of required personnel available for duty, which makes formally measured readiness a function of labor-supply behaviors but not of effort/performance or most types of human capital development.[3] Conceptually, however, we believe that readiness outcomes would be affected by effort/performance and human capital development.

For military organizations not directly related to core combat activities, productivity may be viewed in more traditional, business-oriented terms. The primary objectives of these organizations might include customer service, quality, innovation, cost-consciousness, or other similar concerns. The model applies equally to organizations with these objectives. *Productivity* may be defined as a measure of how well these outcomes are achieved relative to inputs. *Effort/performance* may be defined as behaviors that are instrumental in producing these outcomes.

Personnel Management Processes

Compensation (extrinsic rewards) is, in isolated instances, linked directly to behaviors. More typically, however, rewards are linked to behaviors through one or more processes within the personnel management system. Parts of these processes may be performed by centralized HRM staffs at the service headquarters level, by special field activities such as personnel centers or recruiting commands, by HRM staffs at intermediate and local levels, or by commanders and supervisors.

Accession Screening. Individuals are attracted through recruiting activities to seek membership in a military service—a behavior labeled *joining* in our model. Those

---

[3]SORTS is a system for periodically reporting the personnel, logistics, and training readiness of military units. In SORTS, training readiness can be considered an output measure of some forms of human capital development—specifically, unit training programs.

seeking to join must meet a set of physical, mental, and behavioral standards established by DoD and the services. For most entries, these standards are applied during pre-enlistment processing by the Military Enlistment Processing Command (a DoD activity) and by the services' recruiting commands. For some specialized entry channels, such as service academies or ROTC scholarship programs, the selection process is highly competitive. *Accession screening* is the term we use to refer to the process of selecting entrants from among those interested in joining.

Education/Training Selection. The services offer a variety of opportunities for human capital development through participation in technical training, professional military education, and academic education at service expense. Participation is often selective. For less-expensive or less-exclusive programs, selection may be conducted routinely by commanders, supervisors, training managers, or human resource managers. For more-competitive programs, selection may be conducted by central boards, as for promotion selections.

Evaluation. Behaviors such as effort/performance, human capital development, professionalism at work, responsibility at home, and assignment sorting are evaluated either subjectively or objectively. Almost all service members receive periodic (typically annual or semiannual) subjective performance evaluations rendered by supervisors. Evaluations are typically captured on standardized forms using behaviorally anchored rating scales, often with narrative supplements.[4] Evaluations also take the form of objectively scored tests on general military or occupational subject matter.

Assignment Selection. Reassignment of a military service member (change of job, change of unit, or change of location) can be occasioned by either a "push" or a "pull." A *push* occurs when a service member becomes available for reassignment after completing a fixed-length tour of duty, such as an overseas tour or a lengthy service school. A *pull* occurs when a vacancy must be filled from some pool of eligibles. In some cases, the process may involve negotiations between human resource managers at service and major command levels, current and prospective commanders, and current and prospective supervisors. In other cases, it may be accomplished through automated processes. Individuals are typically given opportunities to record preferences or otherwise interact in the selection process. *Assignment selection* authorities try to fill jobs with individuals who have the requisite training and experience. Moreover, they try to select the best-qualified individuals for the most-demanding jobs.

Labor Supply/Demand Comparison. This personnel management function consists of comparing projected supplies of personnel against requirements at some level of aggregation (typically, grade, occupation, and/or location).

Classification. Upon accession, and at points thereafter, individuals must be screened for entry into or movement between military occupations. *Classification*

---

[4]In practice, many military evaluation systems seem to be characterized by extremely inflated and very undifferentiated ratings, i.e., almost everyone receives a uniformly high rating. Thus, narratives tend to carry more weight than rating scales.

typically involves written tests to measure cognitive aptitudes. It may also involve medical examinations, strength or stamina tests, or more-extensive screening exercises (such as flight screening for pilot-training candidates). Upon accession, the classification process mediates use of enlistment incentives for hard-to-fill skills.

Deployment Selection. *Deployment selection* consists of selecting units or individuals for movement from their home stations to another location for either training or operational reasons. It is often an involved process wherein joint commands levy requirements on service components, which, in turn, levy requirements on intermediate or local commands where unit and/or individual selections are made.

Awards and Decorations. Individuals may be nominated for these uniquely military forms of recognition by commanders, supervisors, or other knowledgeable individuals. Approval may be either local or centralized, depending on the prestige of the award. In some cases, decorations are awarded automatically for service during certain periods or participation in certain campaigns.

Promotion Selection. *Promotion selection* occurs in several ways. For junior officers and enlisted personnel, it may be based simply on a commander's approval after those personnel have served a specified period of time. For middle-grade enlisted members, it may involve a weighted scoring system that combines test results and performance-evaluation indexes (see "Evaluation" above) with other factors, such as time in grade, time in service, and military awards/decorations. For more-senior officers and enlisted members, it usually entails a service-level central selection board that weighs evaluations, assignment patterns, and human capital development. Promotions are often differentiated across occupations, based on labor supply/demand conditions or other considerations that might lead a service to favor one occupation over another.

Retention Screening. Members who indicate by their stay/leave actions that they wish to remain in the service are not automatically afforded the opportunity to do so. In one sense, members are continually screened to ensure they meet military standards. Those failing to meet standards are discharged through administrative or judicial processes. Additionally, enlisted members are screened at each reenlistment point to ensure that they are fit for retention. In some occupations, they may also be required to compete for scarce reenlistment quotas. During force reductions, members may be selected for involuntary separation or retirement by boards that weigh factors similar to those considered in promotion selections. Both enlisted members and officers are subject to up-or-out rules, which require separation or retirement of those not progressing beyond certain grades at certain career points.

Compensation (Extrinsic Rewards)

A third component of our model is a group of compensation elements that provide extrinsic rewards.

Accession Incentives. For enlistments, *incentives* are differentiated by occupation and by measures of recruit quality. Enlistment bonuses are paid to individuals who enlist and complete training in selected military occupations. Additionally, education fund "kickers" (supplements to the Montgomery G.I. Bill education benefits) may also be offered in selected occupations. Service-sponsored pre-commissioning (Academy or ROTC) or medical education can be considered forms of accession incentives. Access to these programs is differentiated by ability and, for ROTC, by academic major.

Basic Pay. Generally the largest element of compensation, *basic pay* is a function of grade and longevity (years of service).

Special Pays. This element includes all forms of compensation that are contingent on the occupation or type of duty performed by an individual. Examples are aviation career incentive pay, submarine duty pay, sea pay, special duty assignment pay, and various medical officer special pays.

Retention and Separation Bonuses. *Retention bonuses* are paid to individuals in designated high-demand specialties who make commitments for specified additional periods of service. Examples include continuation pays for aviators, nuclear-qualified officers, and health professionals, and selective reenlistment bonuses for enlisted personnel. During force reductions, *separation bonuses* have also been paid in designated lower-demand specialties.

Allowances. This element includes various payments intended to offset the cost of feeding and clothing members and of housing members and their dependents. It includes basic allowances for quarters and subsistence or equivalent payments in kind. It also includes uniform allowances, variable housing and overseas housing allowances, cost-of-living allowances, and family separation allowances.

Cash or in-kind compensations in this element are differentiated with respect to dependent status, grade, and geographic location. Single members are entitled to either dormitory rooms or basic allowance for quarters (BAQ) at the without-dependents rate; members with families are entitled to either family housing or basic allowance for quarters at the with-dependents rate.

Travel Entitlements. Individuals who travel from their home installations on military duty or who move from one installation to another are entitled to several kinds of support. Food, lodging, transportation, and movement of household goods and personal effects are typically either reimbursed or provided in kind.

Retired Pay. This element serves as an incentive for individuals to remain in the military workforce until reaching the vesting point, then to leave the workforce or, more typically, to make the transition to nonmilitary employment.

Elements Not Depicted

Generally, the model sought to portray how compensation, mediated by personnel management processes, is linked to behavior. To keep the model focused on these elements, we omitted several forms of compensation that are not typically differentiated with respect to behavior and several forms of rewards that are not typically viewed as compensation.

Noncash medical; morale, welfare, recreation (MWR); and other benefits are not represented in the model because they are not clearly differentiated with respect to behavior. As part of a total compensation package, they do, of course, affect enlistment actions and stay/leave actions.

Time off from work was often cited by subjects in our base visits as an effective and frequently used means of rewarding day-to-day performance. This mechanism seems to work informally and largely without the use of any personnel management system linkages. While recognizing its importance, we did not consider it a form of compensation, strictly defined, and therefore did not depict it in this model.

Various forms of favorable communication may be linked to effort/performance. While they carry symbolic or intrinsic reward value, they may also influence evaluations or find their way into the material weighed by promotion or retention-screening boards. Again, while recognizing their importance, we considered them to be outside of a more strictly defined HRM system and chose not to depict them in the model.

HOW THE THREE COMPONENTS ARE LINKED IN A MILITARY HRM SYSTEM

Figure 2.1 shows how the three components of the HRM system described above—behavior, personnel management, and compensation (and the specific elements within each)—are related in an HRM system. The depicted model reflects a theory—derived from Vroom's (1964) valence/expectancy model and Porter and Lawler's (1968) elaboration of it—of how individual behavior is motivated in an organization.

Following Vroom and as shown in Figure 2.1, we consider individual behaviors to be motivated by *expectancies* about the relationships of behavior, outcomes, and rewards, and the *valence*, or attractiveness of the reward to the individual. This construct clarifies the direction of causality between compensation received and behavior exhibited at an individual level. Since behavior generally occurs before the compensation associated with it, the resulting compensation cannot be said to "cause" the behavior. (An event later in time cannot cause an event earlier in time.) Rather, behavior is caused by the individual's expectancies about his or her behavior and how it relates to compensation.[5] The individual expects that his or her behavior will produce an outcome of value to the organization and that the organization will observe and reward the outcome. Further, the strength of the motivation will be related to the valence of the reward.

Our model depicts a circular framework. Consistent with the discussion above, *expectancies* and *valences* influence *behavior*, which is regulated and evaluated by *personnel management* processes. Decisions emanating from personnel management

---

[5]At an organizational level, causality can be conceptualized in the opposite direction. Individually, behavior causes payment of compensation. Organizationally, a compensation system, if successful, causes desired behavior.

Evaluating the Current Military HRM System 13

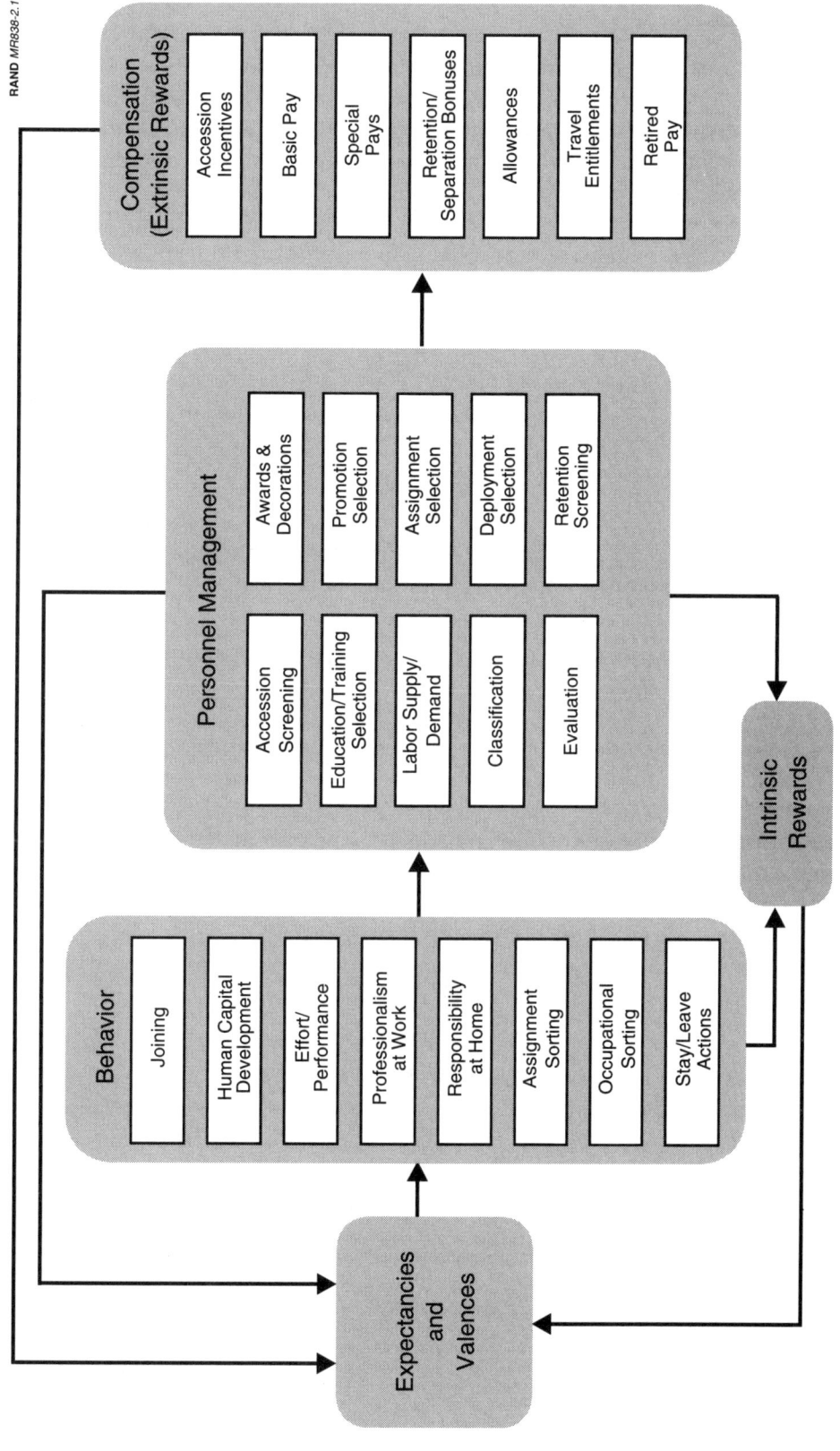

processes then determine the elements and levels of *compensation*, or extrinsic rewards, received by the individual.

Additionally, *intrinsic rewards* can be attached by individuals to the behaviors themselves or to outcomes of personnel management processes. The findings of our base visits support a hypothesis that intrinsic rewards play an important role in shaping the behavior of military personnel. As characterized by subjects participating in our focus groups, these rewards include the inherent satisfaction of serving one's country; performing interesting and exciting missions; operating or maintaining sophisticated, powerful, and expensive equipment; living and working in a variety of interesting geographical and organizational contexts; working as part of a team with like-minded comrades; and steadily advancing to positions of higher responsibility. Many individuals, particularly in the enlisted grades, see themselves as under-rewarded (extrinsically) relative to their levels of effort and relative to their closest referents in the private sector, yet many (especially those in the most demanding fields) also see themselves maximizing their individual efforts.[6] If the data we gathered in our field studies are representative,[7] they attest to the strength of intrinsic, relative to extrinsic, rewards in a military context.

Finally, the individual's expectancies are formed by the ongoing operation of these systems and the organization's apparent intentions about them.

LINKING THE SPECIFIC ELEMENTS WITHIN THE COMPONENTS OF THE MILITARY HRM SYSTEM

In this section, we model the linkages between specific elements of behavior, personnel management, and compensation to provide a view of how the current military HRM system functions to help shape desired behaviors. Many well-established practices and cherished traditions are embedded in the system. To evaluate the displacements of well-established practices and cherished traditions that might result from any changes to the system, it is important to understand the current configuration and its basis. Additionally, the model helps to visualize where the current system may be limited in its support of organizational objectives. The model is detailed because it attempts to depict the military HRM system at a level at which changeable components, and their relationships to desired behaviors, are visible.

In the specific linkages between the elements in the HRM system, depicted in Figure 2.2, all links between behaviors and compensation are indirect, relying on one or more processes within the personnel management system as mediating devices. We discuss important linkages, organized by personnel management process, below.

---

[6]The average level of effort reported by military personnel participating in focus groups we conducted was 56 hours per week.

[7]Since our sample size was small and was not randomly drawn, we cannot make a strong claim of representativeness.

Evaluating the Current Military HRM System 15

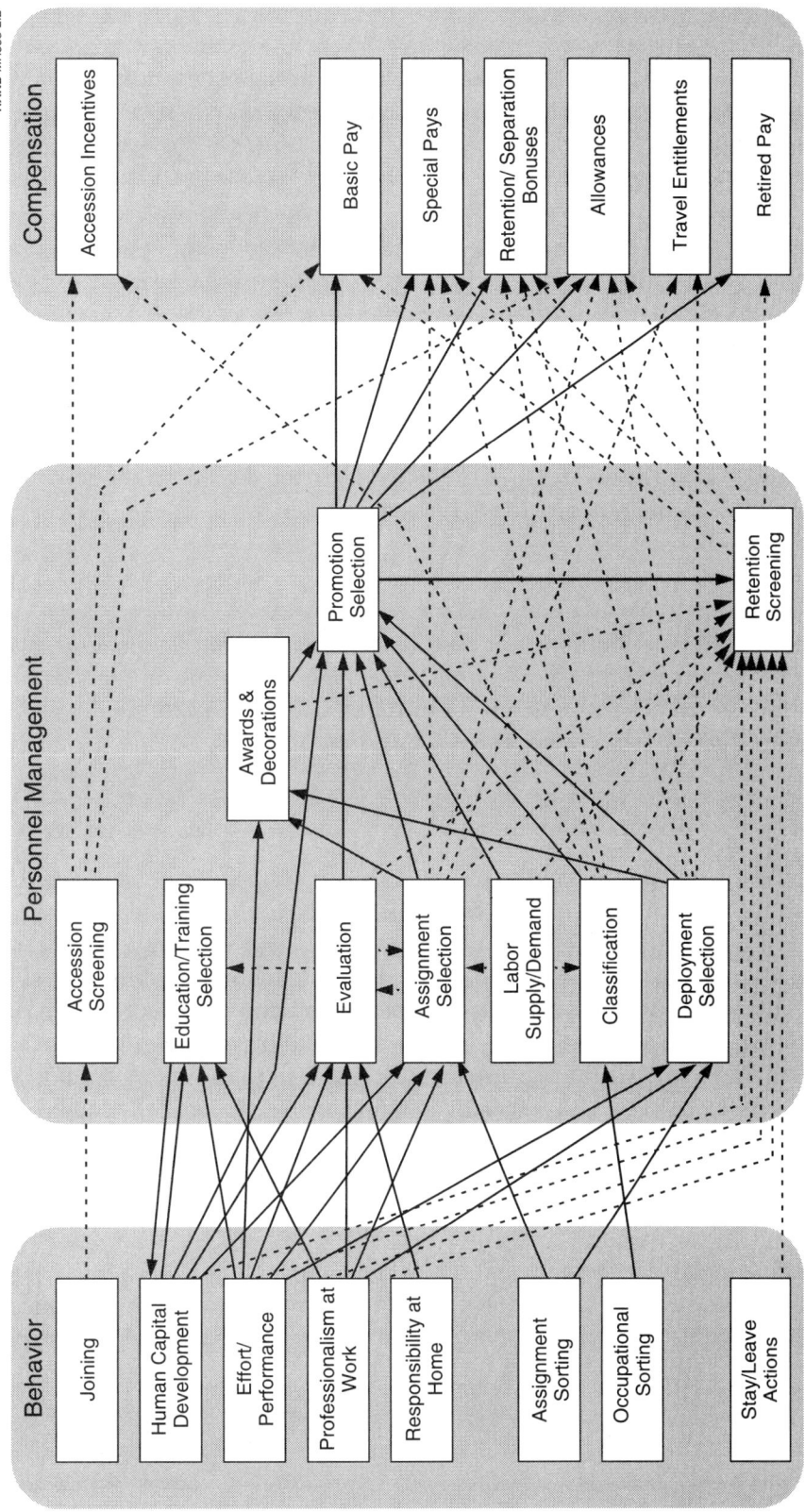

Figure 2.2—Linkages Between Specific Elements in the Military HRM System

NOTE: Solid arrows indicate linkages between behaviors and compensation that are mediated by *promotion selection* (see discussion on pp. 17–18). Dashed arrows indicate other linkages.

Accession Screening

This screening process mediates potential joiners' access to extrinsic rewards such as *basic pay* and *allowances*. Additionally, for individuals qualified and interested in certain hard-to-fill occupations, it mediates access to special *accession incentives*.

Education/Training Selection

Education/training selection seldom or never links directly to a compensation outcome. However, it mediates access to some forms of *human capital development*, which, when subjected to *evaluation* and *promotion-selection*, can lead to greater extrinsic rewards. Individuals may be selected for service-sponsored education or training programs by central boards or HRM staffs, or by local commanders and supervisors. Depending on the program, a variety of behaviors may influence selection, including previous *human capital development, effort/performance*, and *professionalism at work*. These behaviors may influence selection directly, if the selection is done locally, or via *evaluation* processes, if done centrally.

Evaluation

Evaluation does not link directly to compensation outcomes, but is a major factor influencing *promotion selection* and *retention screening*. Depending on service policies, evaluation may be influenced by *human capital development, effort/performance, professionalism at work*, and *responsibility at home*.

Assignment Selection

Assignment selection mediates access to many job-related forms of *special pay* and is an antecedent of *travel entitlements* and one form of *allowance* (variable housing allowance, or VHA). Additionally, it often plays an important role in subsequent *promotion selections* and *retention screening*. While the assignment-selection process is generally administered centrally by service or major command HRM activities, positions that are more critical to the success of an organization are often filled *selectively*, meaning that personnel managers use discretion in nominating candidates and that commanders or supervisors seeking to fill vacancies in selectively staffed organizations have a voice in choosing among available candidates. While authorities may rely in part on formal evaluation tools to make these selections, they may rely as much or more on informal sources of information (personal knowledge of candidates or direct communications with others who have personal knowledge). Past success in competing for challenging and critical assignments is often an important factor in promotion selections or formal retention-screening boards. For advancement to the more-senior grades, officer or enlisted, it may be the strongest discriminator.

Labor Supply/Demand

Labor supply/demand may influence *classification* actions, *assignment selection, promotion selection*, and *retention screening*. The demand side of this process—the number of individuals required in a unit, a military occupation, a grade, or some combination of these—is exogenous to the HRM system. However, the supply side—the number of available individuals—is influenced by a complex web of behaviors and personnel processes. Since these endogenous links are less critical to our use of the model, we do not depict them.

Classification

The *classification* process mediates access to accession incentives for hard-to-fill skills. It is a factor in *special pays* and *retention/separation bonuses*, which are often differentiated across occupations. Finally, it can affect most other elements of compensation through its influence on *promotion selection* and *retention screening*, which are often differentiated occupationally. The process also mediates individual *occupational sorting* behaviors.

Deployment Selection

Deployment selection directly affects *travel entitlements* and *allowances* (specifically basic allowance for subsistence [BAS], which may be reduced during a deployment). It may indirectly affect most other elements in that an individual's history of involvement in deployments can be a factor in *promotion-selection* or *retention-screening* processes. It may be an antecedent for *awards and decorations*, which tend to be distributed more liberally during deployments. Deployment selection influenced by *effort/performance* (depending on how selective the requirement is), *assignment sorting* (if volunteer status is considered), or *professionalism at work* (minimizing personal or family circumstances that impede deployment).

Awards and Decorations

These recognitions do not directly influence any elements of compensation, but are often weighed either objectively or subjectively in *promotion selection* and *retention screening*. They may be influenced either directly by *effort/performance* or indirectly by a variety of behaviors via *assignment-selection* and *deployment-selection* processes.

Promotion Selection

Promotion selection determines grade, which is a factor in most elements of compensation. In most cases, it is influenced by formal *evaluations* and thus by the behaviors that are formally evaluated. Some of these behaviors, most notably *human capital development*, may also be weighed directly in the promotion-scoring system. Because formal evaluations tend to be inflated (most individuals receive uniformly high ratings) and therefore discriminate poorly among promotion eligibles, human capital development evaluated directly in a promotion-selection process is often elevated to the position of a strong discriminator. Promotion selections are often influenced by *classification*, either because *labor supply/demand* conditions are used as a basis to systematically differentiate promotion timing and opportunity across occupational groups or because some specialties (typically, warfighting or other skills closest to the core function of the service) are favored in the promotion-selection process.

Depending on service policies, *promotion selection* may also be influenced by *deployment selection* and *awards and decorations*. Deployments, particularly for participation in real-world contingencies, may be given weight by promotion boards. Awards and decorations can be evaluated by promotion boards or entered directly into weighted-score systems.

Some promotion selections, typically those for very junior grades, may be made by local commanders, who would have the opportunity to weigh behaviors directly or through input from intermediate levels of supervision rather than through the formal *evaluation* process. Since these direct links between behaviors and the promotion-selection process are less typical, we chose not to depict them.

Retention Screening

Retention screening mediates longevity, or years of service (YOS), which, like grade, is a factor in most elements of compensation. Retention screening has many of the same antecedents as promotion selection and assignment selection. Screening may be conducted by a central selection board (e.g., a selective early-retirement board or a continuation board for officers who fail promotion selection), which depends on input from an *evaluation* process. More typically, it is conducted by local commanders and supervisors who have direct or informally communicated knowledge of individual behaviors.

ASSESSMENT OF THE CURRENT MILITARY HRM FRAMEWORK

The existing military HRM framework may not provide sufficient differentiation or appropriate forms of differentiation to adequately value and shape desired traits and behaviors. In some cases, differentiation may be working at cross-purposes with organizational goals, penalizing desired behaviors or rewarding undesired behaviors.

In this section, we examine evidence that changes in the existing system may be warranted. Our insights are drawn from an evaluation of a model of the current system (above); from the perceptions of commanding officers, supervisors, and workers we gathered during visits to military installations; and from a review of HRM practices in other organizations.

Evaluation of the Model

The model depicted in Figure 2.2 illustrates the pivotal roles of promotion selection and retention screening in linking behavior and compensation. Most behaviors influence these processes, and most forms of compensation are affected by them. In the paragraphs that follow, we assess the effects of these multiple linkages of behavior to the promotion-selection process. We do not repeat this assessment for the retention-screening process; however, we recognize that the effects are similar, especially in competitive retention-screening processes such as selective early-retirement boards and selective continuation boards.

We focused on promotion selection because it is the primary mechanism through which extrinsic rewards are differentiated. Our base visits revealed that military members strongly favor using individual performance as a basis for promotion (see Table A.6, Appendix A); they also revealed perceptions among many members that promotion processes often result in poor performers being selected or good performers being passed over.

Specifically, in the focus groups we conducted, military members were asked about their perceptions of the fairness of how rewards are distributed with respect to performance. In many groups, participants cited what they regarded as selection errors in promotion processes (selection of less-deserving people or failure to select more-deserving people). When asked to quantify the extent of the errors, respondents typically estimated between 5 and 50 percent. In written surveys taken by focus-group participants, we also received responses that indirectly confirm a perceived gap between performance and promotion outcomes. When asked whether performance *should* affect pay, 69 percent said yes. When asked if it *does* affect pay, 31 percent said yes (see Table A.5. Appendix A).[8]

These apparent errors could occur simply because evaluation of performance is imperfect. However, the model of the current system shown in Figure 2.2 illustrates that even with perfect evaluation, performance and promotion would probably not be as highly correlated as many seem to prefer. Effort/performance, as measured by periodic fitness or performance reports, is a major element in promotion selections. However, many nonperformance elements, such as occupation, completion of training or education, scores on occupational or general military tests, and longevity, co-determine promotion selections. These elements enter promotion considerations in the form of weighted factors in point-score systems, information that is provided to promotion boards, or the establishment of separate promotion categories with varying promotion rates.

---

[8]Interestingly, the gap between performance and promotion was seen as reflecting the difficulty of measuring performance completely and accurately; most people did not want to have pay based directly on performance because they were skeptical that their performance could be measured objectively. In addition, some expressed concern that pay for performance would undermine intrinsic motivation—the foundation for a norm of excellence. Given these concerns, most were happy to reward performance indirectly through promotion, even though people judged promotion to be imperfectly correlated with performance. They seemed to have some faith that the promotion system tended, in the long run, to screen out poor performers and reward good performers.

Because of these links to disparate behaviors, there is a danger that promotion selection is overused as a mediating mechanism, with two undesirable consequences.

First, it allows individuals to choose their own trade-offs among the types of behavior that influence promotion selections, weakening the capacity of supervisors, commanders, and managers to condition specific behaviors of their choosing. For example, an individual with strong evaluations and a history of assignment selections placing him or her in a series of challenging and critical jobs may perceive an opportunity to reduce personal investment in human capital development, with little or no adverse promotion and compensation consequences.

Second, it makes all relationships between behaviors and consequences less clear. For example, when a member with a strong performance record but undistinguished duty history, low human capital, and peripheral occupation is passed over for promotion, observers unfamiliar with these other discriminators might conclude that promotion is uncorrelated with performance. For these observers, expectancies are weakened and promotion loses some of its capacity to motivate effort/performance.

Promotion selections may also be separated in time from the behaviors they influence. Intervals of five or more years between promotions are not uncommon. During long periods when members are ineligible for promotion consideration or when known probabilities of selection are low, individuals may discount the value of promotion, reducing its valence and, again, weakening its capacity to motivate behavior.

The assignment system's strong role in promotions tends to further weaken the link between documented performance (evaluation) and promotion. The informal, unregulated, and possibly unobserved evaluation systems at work in assignment selection may differentiate between high and low performers more effectively than highly inflated formal evaluation systems. Thus, informal evaluations conducted as part of assignment-selection processes may dominate formal evaluations in their ultimate effects on promotion selections. Although, to the extent that true performance differences are observed and acted upon, informal evaluations can be good for the organization, they can be bad if they are affected by prejudices or forms of networking unrelated to effort/performance or human capital development. In such cases, the system might reward cronyism and other self-promoting behaviors, rather than targeted behaviors.

The behaviors least subject to deliberate shaping through differentiation of compensation appear to be effort/performance and human capital development. Assignment sorting, occupational sorting, and stay/leave actions can each be influenced differentially by pays and bonuses that are specifically linked to them. However, effort/performance and all but a few forms of human capital development are linked to compensation solely through the diffuse, unfocused linkages channeled through promotion selection and retention screening.

Perceptions of Military Members

Our research included visits to one installation from each of the military services, where we conducted focus groups and surveys among military members. The methodology and the pilot data set from our survey questionnaires are reported in Appendix A.

Our field research indicated that the current system of pays, benefits, and other rewards generally takes into account the factors that military personnel think are appropriate.[9] However, we found concerns about several factors, one of which—performance—is discussed above in our assessment of the promotion-selection process. Others are discussed below.

Occupation. The difference of opinion on paying different amounts to different military occupations was strong. On the one hand, many agree that factors such as hazardous duty, time away from family, working conditions, number of hours of work per week, education and skill requirements, and retention difficulties vary from occupation to occupation and that these should affect compensation in some way. On the other hand, many are concerned that paying people according to their occupation sends the message that not all occupations are equally important to military readiness. In addition, by weakening identification with the larger group (e.g., Marines) and strengthening identification with smaller occupational groups (e.g., infantry), some felt that occupational pay differences could reduce the cohesion and cooperation necessary for successful military missions.

Officer Versus Enlisted. Most people favored the present system of two basic pay tables, one for officers and another, less-generous one for enlisted members, on the grounds that officers tended to be more educated and have more responsibility. However, many people complained that junior enlisted members were paid too little for the jobs they perform and that many junior enlisted families experience financial stresses that must be addressed using individual, command, and supervisory resources, at the expense of unit missions.[10] In addition, some felt that it was unfair to pay a first lieutenant (O-2) with 4 years of service more than a master sergeant (E-8) with 20 years of service.

In general, then, there is some consensus that fairness requires an across-the-board pay raise for enlisted members to pay junior enlisted members more in line with the value of their work and their financial needs and to pay senior enlisted members more than junior officers.

---

[9]We heard that, on occasion, desired behaviors might be punished or undesirable behaviors rewarded. As examples of the former, difficult judgment calls about whether to take initiative or follow orders, or whether to be honest or loyal could lead to inappropriate punishment. Also, outstanding performers could be punished, in a sense, by the tendency of supervisors to assign them disproportionately heavy workloads. Focus-group members cited several forms of self-promoting behavior as undesirable but sometimes rewarded, at least in the short run. These include ingratiating behaviors (by subordinates), self-interested leadership (by superiors), and "back-stabbing" (by peers).

[10]Hosek et al. (1992) found, however, that, when controls were established for age, education, and occupation, junior enlisted pay has kept pace with civilian wage growth since 1982. There was no wage gap for this group because, in general, there was little wage growth during this period for young, inexperienced workers with a high school education.

Service. Although people in the focus groups recognized the differences in the nature of the work the individual services perform, no one supported a different compensation package for each service. However, many complained that Air Force personnel enjoyed a better quality of life than those in the other services.

Marital Status. We found a sharp division, including among married people, regarding the present policy of basing allowances on marital status.[11] Critics of the current policy gave three arguments: (1) Pay and benefits should be based on the job you do, not on whether you are married; (2) marriage is a choice, not an entitlement to greater compensation; and (3) the present system creates an economic incentive to get married, which, among junior enlisted personnel, can introduce financial and personal stresses that consume organizational resources.

Practices in Other Organizations

As part of our research, we looked for useful approaches to differentiation that prevail in other organizations but that are not found in the current military HRM system. We identified these practices by reviewing the management literature, conducting several seminars with nationally recognized HRM consultants and practitioners as participants, and by visiting six private- and public-sector organizations known to have innovative HRM programs.[12] The practices we found with potential application in a military context are listed in Table 2.1, with a brief description of each. These practices constitute a portion of the differentiation alternatives discussed in the next chapter.

---

[11]When asked whether marital status *should* affect compensation, our respondents split fairly evenly, with only 55 percent saying it should. However, when asked whether it *does* affect compensation, 85 percent said that it did.

The differences are often substantial. For an E-3 at 2 years of service who marries and moves out of a dormitory, basic allowances for quarters and subsistence totaling about $550 per month at 1995 rates (plus any variable or overseas housing allowance) are added, tax free, to a basic paycheck of $1,050 per month, which is subject to FICA withholding and income taxes. The pay raise appears to be large. Moreover, comparing this additional income to one-half of anticipated household expenses (assuming that the spouse also works and bears half of household expenses), many members might anticipate an increase in discretionary income after marriage.

[12]A useful summary of trends in work-management practices can be found in Cascio (1995).

Table 2.1
Current and Emerging HRM Practices Not Found in Military HRM Systems

Practice	Description
Self-managed work teams	Provides groups of workers the authority to manage their own task and interpersonal processes; holds whole groups rather than group leaders accountable for outcomes
Multipolar performance appraisal[a]	Uses reviews by peers and/or subordinates in addition to reviews by superiors
Competency-based pay	Bases pay on individual human capital rather than on job classification
Merit pay	Makes periodic base-pay adjustments contingent on performance
Performance bonuses	Makes periodic lump-sum payments contingent on performance
Gainsharing	Returns a portion of savings to the workforce when operating costs are reduced because of higher workforce productivity or other efficiencies introduced by workers
Goalsharing	Pays a bonus when members of a team meet established organizational goals; similar to gainsharing, but useful when outcomes other than productivity and efficiency are important

[a]Tables in Appendix A refer to this type of appraisal as a "360-degree evaluation."

Chapter Three

IDENTIFYING AND EVALUATING DIFFERENTIATION ALTERNATIVES

In this chapter, we identify a series of alternatives for potentially improving the current military HRM system. We also recognize that any alternative with attractive features might have undesirable collateral effects. In the first part of the chapter, we focus mainly on potential positive effects—the reasons why military line or human resource (HR) managers might want to consider adopting the alternative. In the second part of the chapter, we evaluate the alternatives more systematically, examining how they affect a variety of interests we think are important to military organizations.

IDENTIFYING THE ALTERNATIVES

From our assessment of the current system in Chapter Two—the evaluation of the model, the results of our base visits to military members, and what we see happening in other innovative organizations—along with the implications of three relevant theories—expectancy theory, social justice theory, and transaction cost economics theory—mentioned in Chapter One, we identified potential differentiation alternatives.

As part of this process, we first identified three major areas in which improvements might be made. Within these three areas, we identified nine potential improvements to which we related 16 policy alternatives. All items are summarized in Table 3.1, along with the sources of research for each (i.e., analysis of the model, theories, base visits). The information in the table is discussed in more detail below.[1]

In developing these alternatives, we relied heavily on the three bodies of theory mentioned above. Readers unfamiliar with the theories who wish to follow these discussions more closely are urged to peruse Appendix B before reading the rest of this chapter.

---

[1]We do not anticipate and would not advocate implementing any of these policy alternatives across-the-board in all DoD activities. Rather, we visualize selective implementation by various communities within DoD, following the *tailored flexibility* approach to HRM design advocated by the Eighth QRMC. In this context, "communities" might be construed to mean services, major commands, or functional areas. Note that the tailored flexibility approach is itself a policy alternative, categorized as a labor-market governance-structure issue.

Table 3.1
Potential Improvements, Alternatives, and Sources of Research

Potential Improvement	Policy Alternatives	Sources of Research
Make Greater or More-Precise Use of Differentiated Rewards		
Increase strength of performance/promotion linkages	• Reduce weight of human capital development in promotion	Expectancy and social justice theories; model analysis
	• Reduce weight of occupational differences in promotion	
	• Increase validity of subjective evaluations through multipolar performance evaluation	
Provide additional performance and productivity incentives	• Establish intragrade merit pay raises	Expectancy theory; practices in other organizations; transaction cost economics theory
	• Establish individual performance bonuses	
	• Establish gainsharing or goalsharing	
	• Introduce self-managed teams	
Increase incentives for human capital development	• Provide competency-based pay for education, fitness, marksmanship, etc.	Expectancy theory; model analysis; practices in other organizations
Further differentiate retention incentives and controls	• Increase special pays/bonuses (to offset less occupational differentiation in promotions)	Social justice theory; model analysis
Permit Greater Differentiation in Labor Markets		
Externalize labor markets	• Relax lateral-entry rules	Transaction cost economics theory; practices in other organizations
	• Increase portability of retirement benefits	
Differentiate labor-market governance structures	• Permit HRM policies and practices to differ across organizations and functional communities	Transaction cost economics theory; practices in other organizations
Eliminate Dysfunctional Differentiation in the Current System		
Eliminate differentiation based on dependency	• Eliminate differences between benefits for members with/without dependents	Social justice and expectancy theories; base visits
Remove mobility disincentives	• Eliminate withholding of BAS from enlisted members who are deployed	Social justice and expectancy theories; base visits
	• Fully compensate members for permanent change of station (PCS) moves	
Resolve perceived enlisted pay fairness issues	• Raise enlisted pay levels	Social justice theory; base visits

Make Greater or More-Precise Use of Differentiated Rewards

In our base visits, we found that current HRM systems give commanders and supervisors limited capability, at least in the short run, to provide high-valence extrinsic rewards for superior performance.[2] Military members like channeling rewards for performance through the promotion system, but our model indicates that effort/performance is only one among many behaviors that influence promotion selections.

Given these circumstances, we saw two possible approaches to improving the reward system. First, the weight of performance in the promotion process could be increased, primarily by reducing the influence of other behaviors. Second, new rewards could be created and could be linked directly to performance or other desired behaviors. These considerations suggest the potential improvements and policy alternatives discussed below.

Increase the Strength of Performance/Promotion Linkages. As discussed in Chapter Two under our evaluation of the current model, the expectancy that links performance and promotion is weakened by the presence of other factors that enter into promotion considerations. Because of these other factors—in particular, human capital development and military occupation—the services sometimes promote weaker performers (who have developed greater human capital or who are in more-favored occupations) or fail to promote stronger performers (who have developed less human capital or who are in less-favored occupations). Removing or reducing the weight of human capital development, occupation, and other factors would strengthen the expectancy that promotion will be associated with better performance. It would also reduce feelings of unfairness in people who believe that promotion should correlate highly with performance but observe instances when it apparently does not.

Reducing the weight of these factors is, of course, not without negative consequences. In addition to serving as the primary means of mediating rewards for past performance (the focus of our attention in this analysis), promotion also serves a very important role as a screening mechanism for placement into positions of greater responsibility. Some forms of human capital development and some occupational experiences may be useful predictors of success in higher-grade positions. Thus, improving the reward-for-performance function of promotions by reducing the weight of these factors might result in selecting less-promising individuals for promotion.

These factors serve other purposes as well. Human capital development contributes to productivity. Occupational differences in promotion selections keep resources aligned with requirements or enhance retention and motivation in favored occupations.[3] If promotion incentives are removed, other incentives might be required in

---

[2] *Valence* is a term used in expectancy theory to denote the power of a reward to influence the behavior of a recipient; it is related to how much value the reward holds for the recipient.

[3] Readers familiar with service promotion systems will recognize that the Air Force differs from the other services in that it has minimal occupational differences in enlisted promotion opportunities. Air Force

order to obtain desired levels of human capital development or retention. Our list of alternatives addresses these needs by including a new mechanism—competency pay—to compensate directly for selected forms of human capital development and increased reliance on selective-retention incentives.

Another way to strengthen the performance/promotion link is to increase the expectancy that superior performance will be observed. Toward this end, the services occasionally adjust their performance-evaluation processes to make them more valid and reliable. We found no fundamentally new approaches elsewhere in the public or private sectors to help the services in refining their traditional supervisory evaluations, but we did see a potentially useful trend toward the alternative of including more perspectives (self, peer, subordinate) in the process. We use the term *multipolar performance evaluation* to describe this alternative.

Provide Additional Performance and Productivity Incentives. Expectancy theory holds that immediate and directly linked rewards provide greater incentives for specific behaviors than remote and indirectly linked rewards. Promotion tends to occur infrequently (after two or three rapid, largely time-based progressions through junior officer or enlisted grades, intervals of three to six years are common); therefore, promotion is perceived to be remotely and indirectly linked to performance. Thus, a more proximate form of extrinsic compensation might be useful in shaping behaviors. We identify three alternatives—merit pay, individual performance bonuses, and team rewards such as gainsharing or goalsharing—commonly used to make portions of pay contingent on performance. With these mechanisms, rewards typically occur immediately following periods of performance of one year or less.

In military HRM, the relationship between supervisor and worker is strictly one to one: Every individual has a designated supervisor to whom he or she is held individually accountable. In the private sector, some firms have found that where individual contributions cannot be easily identified within group or team processes, efficiency is enhanced if the relationship is one to many, with groups rather than individuals held accountable to higher levels of management. These self-managed teams make resource-utilization, process, and HRM decisions that would otherwise be made by a first-line supervisor.

Transaction cost economics provides a framework for interpreting this result. With the entire team held accountable and given governance authority, team members are empowered to observe and deter opportunistic behavior on the part of others within the group and have been given incentives to do so. The team provides a self-enforcing mechanism that may represent the most efficient way for the organization and its employees to relate to each other.

Increase Incentives for Human Capital Development. As discussed above, an extrinsic reward linked directly to targeted forms of human capital development might usefully substitute for promotion as a reward. This mechanism could also apply to important forms of human capital development not typically considered in promo-

---

personnel managers claim that the lack of such differences contributes to a very high level of satisfaction with the enlisted promotion system and does not create unmanageable inventory problems.

tion evaluations, such as high levels of physical fitness among individuals subject to the rigors of combat or board certifications and other special credentialing in professional communities. Expectancy theory predicts that such an approach—a clear line of sight between a behavior, an objectively measured outcome of value to the organization, and a reward—would be effective.

This clear linking of rewards to behavior would be described in transaction cost economics terms as greater ex ante completeness in the employment agreement: Those who satisfy preestablished, objective criteria would get the prescribed rewards, which would be less costly than using the labor-intensive promotion evaluation process (a costly ex post monitoring and enforcement mechanism, in transaction cost terms) needed to reward these behaviors in the current, less-complete employment agreement. More-complete agreements tend to reduce transaction costs when the environment for the transaction is less complex, as appears to be the case here, because the desired forms of human capital development (training, education, physical fitness, physical skills, etc.) in most military environments are probably well known and easily measurable.

Fairness considerations suggest a note of caution. As mentioned above, some individuals, by virtue of environmental conditions—an often-cited example is the difficulty faced by members of frequently deploying units in pursuing advanced academic degrees—have limited opportunities to pursue human capital development and, therefore, less opportunity to earn pay contingent on it. During our base visits, many of our focus-group participants felt that it is unfair to reward a behavior if there is unequal opportunity to engage in the behavior.

Further Differentiate Retention Incentives and Controls. Retention incentives and controls—reenlistment and separation bonuses, reenlistment quotas—are currently differentiated by occupation. We propose further differentiating these incentives and controls so that unbalanced higher-grade manning caused by de-emphasizing occupational differences in promotion selections is offset.

In our base visits with military members, we noted that occupationally differentiated compensation was viewed as unfair by about one-half of the subjects in our sample. When rewards are differentiated, social justice theory predicts that individuals will apply an equity standard to test the fairness of the distribution. This standard is met when outcomes (rewards) are distributed in proportion to recipients' inputs. But military members, for the most part, make little or no contribution of their off-duty time or personal funds to developing their occupational differences. Most enter military service with no specific preparation for their military occupations.

The military services develop occupational differences through the training and experience they provide to their members—at organizational, rather than individual, expense. Thus, members fortunate enough to find themselves in undersupplied skills at reenlistment time are not seen as having *earned* the retention bonuses or more-liberal reenlistment quotas they receive. Because this perception causes dissatisfaction among those not receiving bonuses or facing comparatively tighter reenlistment constraints, we suggest that differentiation of rewards on the basis of occupational labor-market considerations be used sparingly.

Occupationally differentiated rewards are not, of course, based on individual performance or other individual inputs. Rather, they are based on market forces (differing supply and demand across occupations). In our base visits, we encountered members in skills for which supply equaled or exceeded demand (and who were thus not likely to receive retention bonuses). Although these members recognized and accepted the market basis for retention bonuses, they nonetheless did not like it. In the framework of social justice theory, they may recognize that the market basis for occupationally differentiated rewards is procedurally fair, even if it is not distributionally fair.

Changing the Model. These alternatives for making greater or more-precise use of differentiation of rewards would modify the appearance of the model we developed in Chapter Two. Figure 3.1 shows these changes for a subset of behaviors, personnel management processes, and compensation elements that relate to performance management. In this figure, some paths of influence from Figure 2.2 have been enhanced while others have been diminished. New processes and compensation elements not in Figure 2.2 would create new and more-direct paths of influence. The net effect is to take some of the pressure off an overburdened promotion-selection process.

Permit Greater Differentiation in Labor Markets

Transaction cost economics theory holds that governance structures should be differentiated on the basis of differences in characteristics of the working environment, including separability of tasks and the organizational specificity of human capital assets. The internalized labor market that prevails in the military services is at the extreme of a spectrum visualized in this body of theory—efficient for a case of low task separability and high asset specificity but inefficient when these characteristics are absent.[4] At the other end of this spectrum are externalized labor-market arrangements in which organizations establish more-liberal workforce entry and exit rules.

These considerations suggest two opportunities for improvement: (1) externalize labor markets in many functional areas, and (2) vary the internal/external dimension of the labor market, as well as other HRM policies and practices (collectively referred to in transaction cost economics as the *labor-market governance structure*), across organizations or functional areas.

In our base visits, military members frequently cited low task separability and high asset specificity as characteristics of military environments. As the theory predicts for an organization with these characteristics, military organizations have highly internalized labor markets. However, we do not believe these characteristics apply uniformly across organizations and functional areas. In general, we found in our base visits that units further removed from combat tended to exhibit fewer of these characteristics. For example, clerks in a finance office tend to work independently

---

[4]An organization is said to have an *internal* labor market if new employees are brought in primarily at low-level points of entry and most higher-level positions are filled by promotion from within the organization.

Identifying and Evaluating Differentiation Alternatives 31

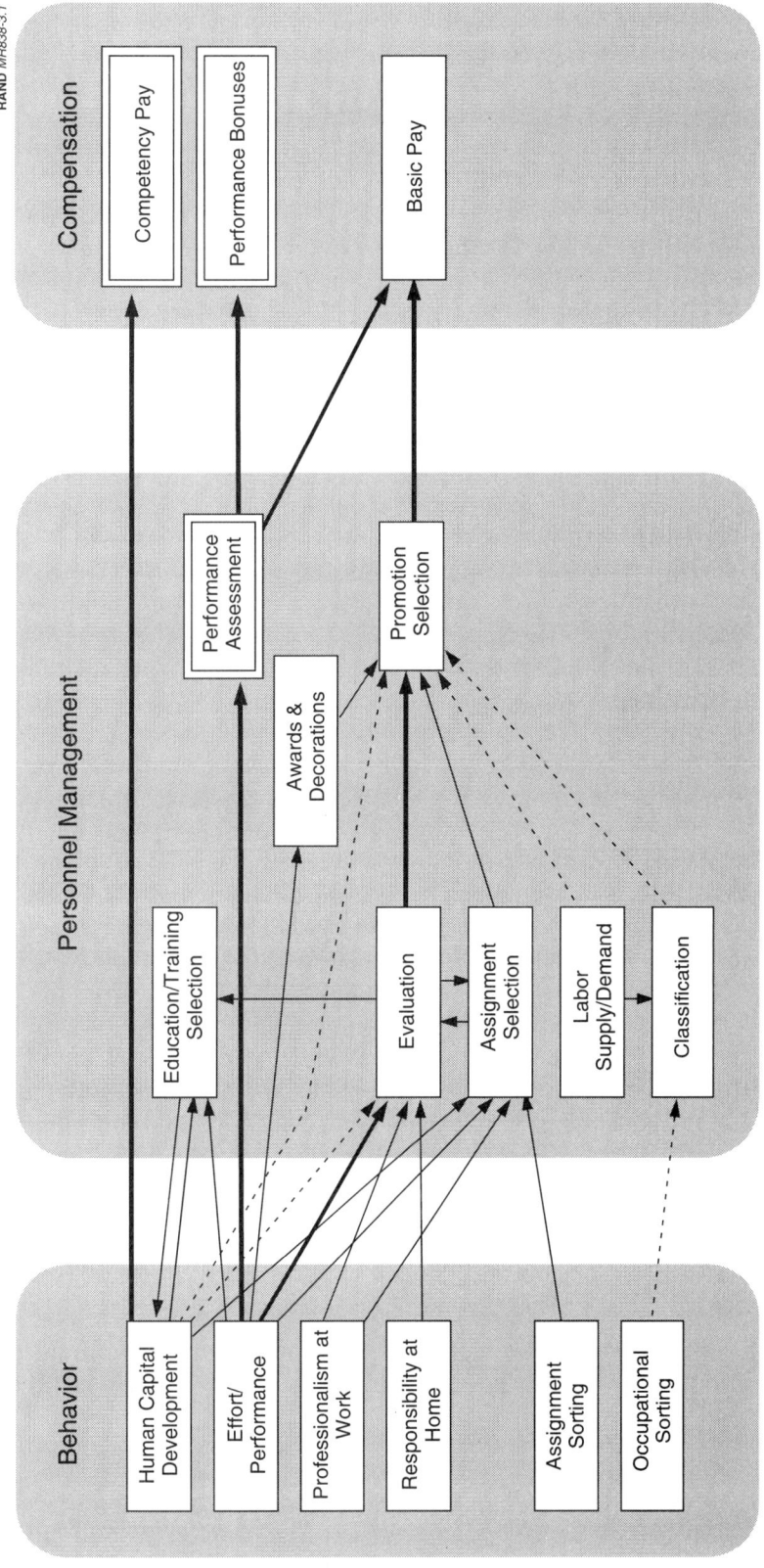

Figure 3.1—Relieving an Overburdened Promotion-Selection Process

NOTE: Plain boxes and lines indicate existing, unchanged elements and paths. Double boxes indicate new elements. Bolder lines indicate new or enhanced paths of influence. Dashed lines indicate diminished paths of influence.

(indicating high task separability) and to do work that is similar to that found in the accounting departments of private-sector firms (indicating low asset specificity). Thus, there may be some opportunity to improve military HRM by varying some provisions across functional areas.

Externalize Labor Markets. In policy terms, *to externalize labor markets* means to relax lateral-entry rules to permit direct appointment of individuals from outside the active-duty military force at grades above E-1 and O-1. When entry constraints are relaxed, advancement opportunities for those already in the organization are adversely affected. To restore balance, members must compete for advancement externally. Thus, if entry constraints are relaxed, there is a corollary need to accommodate freer exits. As a practical matter, this will require making retirement benefits more portable.

Differentiate Labor-Market Governance Structures. In our visits to private-sector firms, we saw a trend toward devolving HRM authority to line managers, movement away from "one-size-fits-all" HRM systems, and a greater tailoring of HRM systems to the varying needs of business units within the firms. If these trends were mirrored in the military environment, services might be permitted to adapt pay tables to their own retention and promotion patterns. Similarly, some functional communities might move toward a greater and more-explicit proportion of pay at risk (i.e., pay that is contingent on individual, group, or organizational performance), more externalization of their labor markets, or different tenure policies.

Any movement along this axis is likely to provoke concerns about fairness. For example, dissatisfaction has been reported as a result of varying service treatment of subsistence, housing, and travel pays for members deployed on joint missions.[5] As with occupationally differentiated rewards, caution must be observed in permitting differentiation along service or functional lines.

We noted in our base visits that people tend to accept what they regard as distributionally unfair outcomes (differentiated, but not in proportion to the recipients' inputs) with minimal apparent dissatisfaction if there are otherwise-valid reasons for them and the distribution is procedurally fair—as with market-based retention bonuses (see "Further Differentiate Retention Incentives and Controls" subsection above). Differentiation of HRM treatments across service or functional communities might be similarly accepted if strong rationales exist and are clearly communicated. For example, an individual bonus scheme in an environment such as recruiting, where individual contributions are critical to the mission and can be measured objectively, is likely to be accepted as appropriate by individuals in other environments where these conditions do not prevail.

We note that special and unique HRM rules have evolved for some communities. Examples include statutory provisions for managing rated officers, officers with ex-

---

[5]Consider, for example, the case of Marine and Air Force personnel billeted at Aviano Air Base, Italy, in support of operations in Bosnia (Anderson, 1995). Air Force officials housed deployed personnel in hotels while Marine officials housed deployed personnel in tents. Service cultures play a role in introducing these differences.

perience in joint activities, and acquisition professionals. Because of their legislative basis, these rules have been applied with some consistency across services. In private-sector examples we have observed, HRM systems tend to be more decentralized and the decision to determine specific features is often devolved to line managers, resulting in much less consistency across business units within a firm, or even across individuals within a business unit.

If the inconsistencies were apparent, social justice theory would predict a greater perception of unfairness among employees. However, private-sector firms seem to counter these perceptions by treating individual compensation agreements and outcomes as confidential information. Since pay confidentiality may be inconsistent with public-sector accountability and freedom-of-information standards, highly systematic approaches to differentiation across functional communities, such as that cited above for rated officers, may be required in the public sector. A high degree of consistency would also have the advantage of making it easier to demonstrate and communicate a rational basis for the differentiation.

Eliminate Dysfunctional Differentiation in the Current System

Our focus-group participants also pointed out several areas in which HRM systems seem to distribute rewards unfairly or to work at cross-purposes with organizational objectives. Opportunities for improvement can be found in eliminating these problems, although, as we note later in evaluating the alternatives, the solutions are often expensive.

Eliminate Differentiation Based on Dependency. Our focus groups and surveys revealed two concerns about the differentiation of compensation based on dependency status. First, as discussed in Chapter Two, there were sharply divided feelings about the fairness of this differentiation. Second, there were concerns that the differentiation might be great enough to motivate a higher rate of family formation among junior enlisted members.

Social justice theory provides conflicting predictions about how people will react to this form of differentiation, which violates equity and equality standards but conforms to a need standard.[6] Some members consider the need standard appropriate; others see equity or equality standards as being more appropriate. All standards would apparently be met simultaneously if the pay of all members were elevated to the level of greatest need (current with-dependents rates), but the cost would be high.

Remove Mobility Disincentives. Military duty often requires deployment and/or relocation of forces or individuals to where they are needed. Ideally, members willingly respond to these needs. However, military members have a unique and legally

---

[6]An *equity standard* is met when rewards are consistent with individual inputs or the results they produce. An *equality standard* is met when rewards are equal regardless of contributions or results. A *need standard* is met when rewards allow individuals to meet living requirements appropriate to their station in life. These concepts are described more extensively in Appendix B.

binding obligation to go when ordered, even if they are unwilling. Under some circumstances, military members are penalized economically for these highly desired mobility behaviors. The paragraphs that follow describe two such disincentives and their unfavorable behavioral implications.

When enlisted members are deployed for field or sea duty, they are required by law (37 USC 402) under most circumstances to forgo basic allowances for subsistence. Although the enlisted BAS for those granted permission to mess separately is $6.98 per day (1995 rate), or over $200 per month, it is unlikely that typical household grocery bills for married members decline by that amount when they are deployed. Thus, their families are less well off during deployments.

In reaction to widespread dissatisfaction, this disincentive was recently eliminated, in the highly visible actual contingencies of Somalia, Haiti, and Bosnia, by defining *operational missions* to be other than field duty. Members deployed for these contingencies were thus allowed to continue receiving BAS. However, for routine training and administrative deployments, the disincentive still exists. In our base visits, some supervisors cited this disincentive as contributing to opportunistic behaviors whereby individuals render themselves unavailable or ineligible for deployment.

When members are reassigned to a new permanent change of station (PCS), they receive allowances or services in kind to offset the costs of relocating themselves and their families. Many members find these benefits inadequate to offset the full costs of disestablishing and establishing households. Moreover, military relocation benefits are noticeably less than benefits provided to relocating federal civil service employees. For example, civil service employees are entitled to reimbursement for real estate and lease transaction costs, whereas military members are not. The services generally seek to select qualified volunteers over nonvolunteers when relocations are necessary. By decreasing the pool of volunteers to move, this economic penalty probably interferes with the best match of people to positions.

Given that voluntary, or at least more readily accepted, mobility is valuable to military organizations in the two situations described above, it would be better obtained by eliminating economic penalties that are imposed in a differentiated way: The penalties rise in proportion to the degree to which members face mobility requirements. To avoid the penalties, policies (and their statutory basis) must be changed to eliminate withholding BAS from enlisted members who are deployed and to provide relocation benefits that fully compensate members for PCS moves.

Resolve Perceived Enlisted Pay Fairness Issues. In our focus groups, we discovered two widely held fairness concerns related to enlisted pay. These concerns touched both ends of the seniority spectrum.

First, both junior enlisted personnel and their supervisors were concerned about junior enlisted pay levels. Junior enlisted members, especially those with families, often experience financial stress. This stress reportedly detracts from their performance of military duties. Moreover, supervisors reportedly invest significant effort in dealing with the effects of low pay levels, both by counseling junior members to help overcome family-support problems and by taking disciplinary action to deal with financial irresponsibility. In the framework of social justice theory, the part of this financial stress that is not attributable to irresponsible spending is a sign that junior enlisted pay violates a need standard.

Second, both officers and enlisted personnel perceived that senior enlisted members are underpaid relative to junior officers.[7] Senior enlisted personnel are often as well or better educated than junior officers, have more experience, and are often seen as contributing more to the effectiveness of their units. Since the relative compensation level of senior enlisted members does not seem proportional to their relative inputs, it is seen as violating an equity standard.

Since the enlisted pay scale seems low at both ends, although for different reasons, an apparent alternative would be to raise enlisted pay across the board.

EVALUATING THE ALTERNATIVES

In this section, we evaluate the 16 policy alternatives discussed above and summarized in Table 3.1. We first describe and provide the rationale for the organizational interests we examined in this evaluation. Then, we predict the effects of the alternatives on these interests. We base our predictions on applying relevant theory to the cases at hand and on factoring in feedback to some of these alternatives we received from military members in our base visits.[8]

Rigorous empirical testing of these predictions was beyond the scope of our research. While some empirical evidence is available from other studies to support some assessments, it must be applied cautiously in predicting results, in a military context, which could differ from other contexts in important ways. Lacking empirical support, these predictions are generally insufficient in themselves to support policy formulation, but they should help policymakers in the Office of the Secretary of Defense (OSD) determine what additional HRM flexibilities ought to be considered and help policymakers in various communities within DoD match alternatives to their varying needs. In general, for alternatives that require a significant shift from past military practice, relevant empirical evidence must be developed through carefully designed, context-specific testing.

Finally, with a view toward possible implementation, we categorize the alternatives along the dimensions of cost and need for statutory or policy adjustments.

---

[7]There is a limited range of overlap between officer and enlisted pay tables. The basic pay of an O-3 with 4 years of service exceeds the basic pay of all enlisted personnel except E-8s with 26 years and E-9s with 22 years of service. An O-2 with 2 YOS is paid about the same as an E-7 with 14 YOS.

[8]We do not believe that the reactions of military members should govern acceptance or rejection of alternatives. To some degree, these reactions are the products of socialization and greater comfort with what is familiar. However, the reactions do highlight concerns that must be addressed if any of the alternatives are to be incorporated in an HRM design, and they suggest the magnitude of member involvement, education, and information-sharing that may be required as part of design and implementation efforts.

Organizational Interests

Our own research and our discussions with the staff of the Eighth QRMC suggested a range of organizational interests that might be affected by shifts in HRM policies or practices. We list these interests below, indicating why we include them and describing our general approach to predicting how they will be affected.

Productivity. Any organization with strategic objectives envisions organizational outputs or outcomes necessary to achieve them. *Productivity* is a measure of the efficiency with which the organization employs its resources, including human resources, in producing outputs or outcomes. *Outputs* can include "widgets," services, or intangible products such as ideas. *Outcomes* include states or conditions that organizations attain through their activities. For example, a military organization might seek readiness as an outcome while a private-sector firm might seek market share. One important characteristic of a strategically aligned HRM system is its contribution to producing the desired output or outcome efficiently. In the private sector, increasing this efficiency contributes to profitability. In the public sector, it lowers resource demands.

HRM systems can affect productivity in a variety of ways. They can favorably influence productivity by, for example, increasing the motivation of employees, encouraging human capital formation, or attracting and retaining the right kinds of employees. As we see below, some alternatives can increase productivity through one of these channels while decreasing it through another.

Compatibility with Current and Future U.S. HRM Practices. The degree to which military HRM practices are consistent with practices in other public- and private-sector organizations is important in several respects. This consistency has been cited as a factor supporting public confidence in the military as an institution (Thie and Brown et al., 1994).[9] Moreover, sharp differences can add friction to the movement of workers between military and nonmilitary workforces by imposing costs on individuals making these transitions.

Organizations may have valid strategic reasons for developing an internal labor market (Williamson, 1975), characterized by intentional impediments to entry or exit of experienced employees (Bridges and Villemez, 1994). However, an organization that wants to strategically align its HRM system must be free to move toward greater reliance on an external labor market. If incompatibility of military and nonmilitary HRM systems creates impediments to entering or exiting military service at other than traditional, well-defined points, the option of greater reliance on external markets will not be available to military organizations.

Fit to Current and Future Military Organizational Structures. Some features of HRM policies and practices may directly influence strategic objectives and therefore must be congruent with an organization's choice of organizational structures. An organization would not, for example, want to emphasize individual rewards if its pro-

---

[9]Although Thie and Brown et al. (1994) provide a rational basis for this claim, it has not been tested empirically.

cesses were team-oriented, or team rewards if its processes consisted of independent tasks (Wageman, 1995). Thus, alternatives for military HRM systems can be evaluated in terms of how well they support current or anticipated forms of military organizational structure.

Our assessments of this fit were, of necessity, conditioned by our expectations and assumptions about military organization structures, which we summarize as follows:

- At operating levels, combat-oriented military organizations will remain essentially hierarchical. Positive command and control of large, powerful forces demands this form of organization. Nonetheless, hierarchies may be flattened to some degree, echoing widely observed, efficiency-motivated trends in the private sector.

- Jointness will continue to grow in importance. The 1986 Department of Defense Reorganization Act (the Goldwater-Nichols Act) provided a strong impetus. The increasing voice of the Joint Requirements Oversight Council (JROC) in weapon-acquisition and defense-budget deliberations provides more impetus (Blaker, 1996), as does the visible role played by joint commanders and joint headquarters functions in planning and implementing major contingencies, such as the Persian Gulf War and operations in Somalia, Haiti, and Bosnia. Joint operations typically comprise units reporting administratively to service chiefs but operationally to joint commanders. This well-accepted arrangement is a form of matrix organization and a departure from a strictly hierarchical form.

- Boundaries between civilian and military functions will blur and shift. This trend can be seen in calls for privatization of functions performed, in some circumstances, by military members (White et al., 1995) and by the increasing presence of civil service and contractor personnel in the vicinity of battlefields. These individuals are not subject to characteristic military discipline and accountability structures such as the Uniform Code of Military Justice (UCMJ). Thus, commanders in the field have less authority over civilians and contractors than they do over military personnel.

- As information-handling technologies and competencies continue to develop (Toffler and Toffler, 1993), organizations will rely more on networking and less on hierarchical relationships to coordinate activities.

These expectations indicate movement toward military organizations that remain hierarchical, but less consistently so than they have been. In assessing the effect of HRM alternatives on military organizations, we find it necessary to recognize both traditional needs for hierarchical control of military forces and emerging needs for organizational flexibility.

Fit to Military Cultures. Organizational cultures help people cope with uncertainties by providing them with accepted ways of expressing and affirming their beliefs, values, and norms (Trice and Beyer, 1993). HRM policies and practices are shaped by and also help to shape organizational cultures. Thus, a "fit," in this context, depends on either the extent to which an alternative is compatible with a valued aspect of an organization's culture or the extent to which it promotes a desired change in the culture. In assessing this fit, we recognize that, at one level, all military service occurs in a common cultural context that differentiates it from civilian employment. At another level, cultures differ significantly across services, functional communities, types of military units, and level of seniority.

The common culture of military service stresses values such as service, duty, patriotism, integrity, and trust. Military members come to see themselves as members of an institution that has important and noble purposes for which they will, if necessary, sacrifice their personal interests.

Differences among the cultures of the military service have been described by Builder (1989) and other observers. We also observed differences among the services in our base visits. We can visualize even more striking differences between the cultures of the military services and those of the nondefense uniformed services (United States Coast Guard, National Oceanic and Atmospheric Administration, and Public Health Service). The OSD and the defense agencies have a culture distinct from that of the services—more civilian than military and more management-oriented than operations-oriented. Joint organizations introduce yet another cultural contrast.

Cultural differences would also be expected across functional communities. We would expect to find differences in values and behavioral preferences in, for example, operations as opposed to maintenance communities, education as opposed to training communities, combat as opposed to combat support communities, and so forth.

In our focus groups and surveys, we expected but actually found very limited differences in behavioral preferences across functional communities. (See Appendix A, Tables A.7–A.13.) The differences we found were primarily between services. Somewhat fewer differences were found between officer and enlisted members, with almost no differences between combat and noncombat communities or junior and senior personnel. We suspect that there are differences we failed to detect, either because our respondents lacked the vocabulary to articulate the differences or because our methodology was not penetrating enough.

Fairness. Research on organizational justice (see the discussion of social justice theory in Appendix B) suggests that people who perceive themselves to be unfairly treated take actions that have major consequences—often negative—for themselves and their organizations (Sheppard, Lewicki, and Minton, 1992). Perceived unfairness can cause loss of confidence, loss of commitment, loss of support, and, ultimately, loss of effectiveness.

HRM policies and practices govern much of the treatment of people by their organizations and are thus particularly likely to be judged against standards of fairness. Fairness, however, can be difficult to assess because of conflicts between equity, equality, and need standards. For each alternative, we first assess which standard or combination of standards is appropriate—*equity* where performance is a goal, *equality* where a sense of community is an important consideration, and *need* where human dignity is involved. We then use each applicable standard to assess its effect on the organization.

Accountability. In a traditional bureaucratic organizational setting, individuals are held accountable for meeting organizational standards and contributing to organizational effectiveness primarily through a hierarchical pattern of supervisory and staff oversight. However, many contemporary HRM innovations entail shifts of authority from higher to lower organizational levels (decentralization, empowerment, networking) and from central personnel management staffs to line managers (devolution). As these shifts occur, bureaucratic patterns of accountability are diluted; greater reliance must be placed on nonhierarchical legal, professional, and political patterns. The availability and suitability of these other patterns of accountability thus become factors to be considered in evaluating HRM innovations. Moreover, cultivation of other patterns may be an important part of an implementation strategy, leading to long-term acceptance of new management flexibilities (Romzek and Dubnick, 1994).

Appendix C contains an expanded perspective on the issue of public-sector accountability, including a review of the state of nonhierarchical patterns in DoD. Nonhierarchical patterns are based on legal and inspectors general systems, accounting and information systems, political and public-interest mechanisms, and professional standards. In our assessment, we noted weaknesses in the DoD financial-management system and other such circumstances that limit DoD's current capacity to shift to nonhierarchical patterns.

We would advocate energetic efforts to strengthen nonhierarchical patterns, but we do not think that they can provide effective accountability in their current state. Further, we note that HRM systems have the capacity to strengthen or weaken DoD's only currently robust pattern of accountability—hierarchical oversight—by giving supervisors greater or lesser voice in HRM outcomes. However, HRM systems have little or no capacity to affect nonhierarchical patterns. Thus, we favorably assess any alternative that strengthens accountability by reinforcing hierarchical oversight and unfavorably assess any alternatives that weaken oversight or increase reliance on poorly functioning nonhierarchical patterns.

Cost. Given the project's scope, we made no attempt to assess costs other than to describe likely increases or decreases at a very general level. We assess a moderately increased cost for any alternative that introduces incentive pays. If these pays are added to existing pay elements, costs go up. If they place normal pay entitlements at risk, we assume that a risk premium will be paid, which also increases costs somewhat.[10] We assess a budget-neutral effect for alternatives that can be funded through offsets and that do not require a risk premium. We assess heavier cost burdens for any alternatives that are likely to entail significant administrative costs (such as increased demands on supervisors) or that call for pay increases that cannot be readily offset.

We recognize that some alternatives might generate productivity gains that more than offset additional compensation costs, yielding a potential for a positive effect

---

[10] For a discussion of how increased risk is associated with shifts to output or outcome measures of performance, see the final paragraph in Appendix C.

(reduced cost). However, our cost and productivity assessments lack the precision needed for this kind of cost/benefit analysis. Similarly, any alternative that shifts retention would have cost implications, such as changes in needed retention bonuses or changes in longevity, that could make the force more or less expensive. Again, our study did not provide the tools to assess these effects. We limit our assessment to first-order cost effects only.

Organizational Interests Not Assessed. Some evaluations of HRM system outcomes have focused on interests we have chosen not to address. Notably, we have not systematically assessed net changes in retention, job satisfaction, and racial or gender equal-opportunity issues.

Retention and job satisfaction are generally net outcomes influenced by many different factors. With sufficient empirical data, they can be evaluated in reduced form. However, we lacked the necessary empirical data for this kind of analysis (and regarded gathering it to be outside the scope of this project). Elaborate theoretical models can also be constructed to link HRM alternatives to these outcomes.

Employee turnover, for example, has been linked to reduced levels of job satisfaction and organizational commitment (Steers and Mowday, 1981). We expect retention to be influenced by fairness, which probably contributes to job satisfaction, and a shared organizational culture, which probably contributes to organizational commitment. We also regard this more-elaborate modeling to be beyond the scope of our current research, limiting our observations about retention or job satisfaction to simple first-order effects (e.g., effects of adjustments to retention bonuses).

Since greater differentiation might influence racial or gender equality, we might have assessed this as a separate and specific interest. We chose not to do so on the assumption that norms for procedural fairness in military organizations would generally guard against discrimination or favoritism. We recognize that these norms will occasionally break down in practice, but we did not develop insights into when such a breakdown might occur or how our alternatives might influence the propensity for it to occur.

Assessing Implications of the Alternatives for Organizational Interests

We assessed the implications of each policy alternative for any organizational interests affected by the policy.

Reduce the Weight of Human Capital Development in Promotion. Promotion selection serves as a reward for past effort/performance, a reward for other desired behaviors (such as human capital development), and as a means of screening for assignment to positions of increased responsibility.

When promotion is evaluated as a reward mechanism, reducing the weight of human capital development has both positive and negative productivity implications. On the one hand, it can improve productivity by improving the line of sight between effort/performance and reward. When the weight of human capital in promotion considerations is reduced, the weight of the remaining factors, including performance, is increased. When performance carries more weight in promotion considerations, members will likely develop a stronger expectation that performance will be rewarded through promotion. On the other hand, this can reduce productivity in the long run by removing an incentive for human capital development.[11]

By strengthening the voice of superiors (who provide performance evaluations) in determining promotion outcomes, this alternative enhances the fit to hierarchical military organizational structures and accountability. And fairness, or equity, is improved when proportionality of effort/performance to rewards is increased.

When promotion is evaluated as a screening mechanism, the results are less favorable. Future productivity declines, because individuals with less human capital may be promoted over individuals with more human capital. Perceptions of fairness diminish for individuals who believe that human capital development should affect promotion outcomes.

In de-emphasizing human capital development, the services might proceed by first differentiating among its various forms. Past performance in key positions, for example, is more critical to success in positions of higher responsibility than are other forms, such as physical fitness; therefore, it is important to the screening function of the promotion system. Some forms, such as professional military education, are more universally accessible than others and, therefore, are more compatible with a fairness standard for the promotion system. Some, such as physical fitness or marksmanship, are more perishable than others and, therefore, are incompatible with a long-term reward mechanism such as promotion. Retention of only highly critical, widely accessible, and less perishable forms of human capital in promotion considerations would appear appropriate.

Reduce the Weight of Occupational Differences in Promotion. When promotion is evaluated as a reward mechanism, this alternative has positive implications for productivity, hierarchical organizational structures, equity, and accountability similar to those of reducing the weight of human capital development in promotion considerations. On the negative side, it would diminish the capacity of the services to favor core functions or occupations in promotion selections and, thereby, to differentiate the cultures of their units or functions. Organizations may want to cultivate competencies that are closely associated with their core functions—in military organizations, warfighting—and that are not evenly distributed across occupations.

When promotion is evaluated as a screening mechanism, a negative effect on productivity would be expected if it is true that productivity is enhanced either by matching grade inventories to grade requirements in an occupation or by favoring critical occupations in promotion considerations. Favoring certain occupations in the promotion process influences the composition of the population from which future leadership is drawn. However, promotion advantages based on occupation are usually seen as unearned, resulting from labor supply and demand rather than from

---

[11]By definition, *human capital development* includes any training, education, experience, or other form of development that leads to higher productivity.

merit and, therefore, not meeting an equity standard of fairness. In our surveys of focus-group participants, only 57 percent of respondents thought that occupation should affect compensation. Occupational differences can detract from the cohesion resulting from a shared military culture. Therefore, occupational differences in promotion rates might best be limited to instances in which they are clearly a factor in succession planning or in which the organizational benefits of favoring certain skills are clear and compelling.

Increase Validity of Subjective Evaluations Through Multipolar Performance Evaluation. This alternative would likely combine many positive and negative results. It could increase productivity by better identifying and rewarding stronger performers. It would match a trend found in innovative private-sector firms. It could undermine hierarchical organizational structures by causing individuals to curry favor among subordinates, but could be useful where flexible organizational structures and decentralized decisionmaking might be found advantageous (probably outside of core combat environments). It might have negative implications for a shared military culture, because it could undermine loyalty and responsiveness to hierarchies. But it could help to differentiate combat from noncombat cultures.

Since performance appraisal clearly serves a performance goal, an equity standard of fairness is appropriate. Equity is enhanced if additional observers help to eliminate rater error, but is reduced if peers do not fairly rate others with whom they are competing for promotion or other rewards. Accountability would generally be enhanced because individuals would be subject to additional forms of scrutiny. Administrative costs of soliciting, completing, collecting, and synthesizing multiple inputs would be high.

In our focus groups, people in noncombat environments were more sensitive to some of the positive effects (a more accurate assessment of performance and a resulting increase in productivity); people in combat environments were more sensitive to some of the negative effects (high administrative cost and undermining of hierarchical authority). However, both groups considered multipolar performance evaluation mildly unappealing. People seemed to think that peer evaluation would become "spear evaluation," because peers are often in direct competition for promotions and job assignments, and would be unfairly critical of each other. People did not like subordinate evaluation because they believed it would become a popularity contest and because subordinates seldom see the "big picture" clearly enough to understand why their superiors make the decisions they do. Also, we found more tolerance for its use as a developmental tool (helping individuals understand their strengths and weaknesses) and less tolerance for linking it to competitive processes for promotion or other performance rewards.

Establish Intragrade Merit Pay and Individual Performance Bonuses. Theory predicts productivity gains as resulting from more-direct forms of extrinsic rewards for performance. However, our field research indicates strong misgivings among many service members. Intrinsic rewards seem to play a significant role in military retention and motivation. Attempts to link extrinsic rewards more directly to performance can potentially undermine intrinsic rewards by suggesting that the demands of military service are fully compensable in cash. Additionally, fairness concerns were raised by both senior and junior members, who were skeptical that the best performers can be successfully identified in the short run, especially when supervisors must rely on subjective evaluations of performance rather than objective measures.

Tolerance for greater use of extrinsic rewards was generally low in core combat communities but increased in communities that were more removed from combat, perhaps because tasks are more separable in noncombat communities. Overall, our survey of military members indicated a mildly favorable response to pay for individual performance.

These two alternatives have the advantage of matching common private-sector practice. If implemented in selected communities, they could undermine the shared culture of military service but could help to create differentiated cultures in those communities.

Between the two mechanisms, bonuses seem to have several advantages over merit pay raises. First, because bonuses are one-time rewards rather than additive, the result of a bad decision (from incomplete information, for example) is less enduring. Second, bonuses should enhance performance more than merit pay raises of equal present value because merit pay raises, spread out over time, are discounted more steeply by individual recipients than by the organizations awarding them.[12] Finally, a merit pay raise plan could impose heavy administrative costs: a change to the basic pay table and, in its most complete form, administrative consideration of each individual's annual salary adjustment. Payroll costs for either mechanism can be offset by across-the-board decrements in other pay accounts, but a risk premium would be required.

Establish Gainsharing/Goalsharing. Gainsharing and goalsharing have the potential to align group behavior with organizational objectives, thereby increasing productivity. However, each alternative poses difficulty for attaining this alignment in military environments.

In a gainsharing plan, workers share in input cost savings, subject to acceptable outcomes. In conventional military organizations, the most important peacetime outcome is readiness. However, most aspects of readiness are measured in terms of inputs (people, equipment, training) rather than in terms of an ultimate outcome—a summary measure of potential combat effectiveness. Since a true outcome measure is not available, the consequences of economizing on inputs cannot be easily evaluated. We believe that, lacking an output measure, readiness would be placed in jeopardy if core combat units or the logistics units that deploy with them were given an incentive to underspend budgeted inputs. Gainsharing has its clearest applica-

---

[12]Military members are thought to have mean real personal discount rates in the range of 9 to 14 percent (Black, 1983), while military organizations would discount future streams of benefits at the government's real cost of capital, or about 2 to 3 percent. Thus, if a merit pay raise and a lump-sum bonus had equal present values from the perspective of a military service, the merit pay raise would appear less valuable to a member.

tion in production functions, such as maintenance depots, where outputs are readily measurable and where productivity clearly outweighs environmental factors in driving costs.

Goalsharing awards, to be viewed as fair elements of compensation, must relate to outcomes over which responsible groups have control. Military core combat units generally have well-defined training, logistics, and other readiness-related objectives, but these are often interrupted by real-world contingencies or are influenced by intentionally making some units robust at the expense of others. Logistics units are, in turn, affected by contingencies involving the units they support. To keep a goalsharing arrangement fair, much ad hoc adjustment of goals might be required to make allowances for contingencies or uneven resourcing. Real and perceived fairness might be difficult to attain.

Outside core combat and related communities, conditions may be more amenable to gainsharing and goalsharing. For gainsharing, however, unit operating costs are typically difficult to determine in defense accounting systems, in which many expenses (including military personnel costs) are either centralized or are not itemized and can only be imperfectly allocated.[13] Additionally, setting performance benchmarks (e.g., readiness)—an essential ingredient in gainsharing plans—would be difficult. Goalsharing, too, depends on the ability to measure meaningful unit outputs or outcomes.

Aside from productivity, other results are mixed. Gainsharing is an established private-sector practice, and goalsharing is emerging as a recognized practice in both public and private sectors. Their use in military applications would contribute to compatibility with overall U.S. HRM practices. Their increased emphasis on extrinsic rewards could undermine a cultural emphasis on intrinsic rewards, but could be useful in differentiating cultures. Since a performance goal applies, equity is the appropriate standard of fairness; providing equity is highly dependent on the ability to measure costs (gainsharing) or performance outcomes (goalsharing). Of the people we encountered in our base visits and firm visits, and those who have evaluated DoD financial management systems, many are skeptical that such measurement can be done well in military environments. Additionally, in military environments, costs and other outcomes are frequently subject to environmental shocks (e.g., unanticipated deployment) beyond the control of groups whose rewards are made contingent on those costs or outcomes.

Accountability is degraded if these schemes rely on inadequate accounting and information-management systems to provide objective bases for differentiated rewards. Measurement requirements impose some administrative costs, but bonus pools can be offset by across-the-board decrements in other pay accounts (with the addition of a risk premium).

Our survey of military members showed that overall reaction to gainsharing and goalsharing is neither favorable nor unfavorable.

---

[13]A more extensive discussion of defense accounting system shortcomings is provided in Appendix C.

Introduce Self-Managed Teams. The essential characteristic of a self-managed team is holding a whole team, rather than the leader of a team, accountable for its outcomes. Introducing this arrangement in military environments could bring the same advantages that have made it attractive to some private-sector firms: greater worker investment and involvement in work-process improvement and in product quality. To be viable, however, the arrangement might require a number of conditions not often found in military units, including well-defined outcomes, stable groups, a willingness to provide teams with significant discretion over use of resources, sufficient time for group decision processes to play out, and alternatives to hierarchical patterns of accountability.

Because it undermines the role of rank and precedence, we do not believe this alternative, in its pure form, will find acceptance in combat and related communities. Applications might be found in noncombat communities, although even here our base visits revealed concerns about how hierarchy would be affected and about the equity of rewarding strong or weak team members on the basis of team outcomes. Overall reaction, as measured by our survey, was neither favorable nor unfavorable.

Provide Competency-Based Pay. Competency-based pay for human capital development would enhance productivity by increasing development of needed human capital. It would increase compatibility with other public- and private-sector practices, where competency-based pay is gaining favor. It could reinforce behaviors that are valued across shared military cultures, such as development of high levels of fitness and marksmanship, or more job-specific behaviors that help to differentiate cultures. It would meet a fairness standard, because rewards would be equitably distributed in proportion to individual investment in human capital. However, a negative element would also be related to fairness, because access to human capital development opportunities is uneven. Its costs could be funded through across-the-board decrements in other pay accounts.

When tested in our base visits, this alternative received slightly favorable responses. It could be especially useful for desired forms of human capital development that do not warrant inclusion in promotion considerations.

Increase Special Pays/Bonuses. Retention bonuses, incentive pays, and reenlistment quotas help to align supply and demand of experienced personnel, thus enhancing productivity. Some forms, such as flight pay and submarine duty pay, also contribute to differentiating among cultures within the services. Similar to differentiation of promotion selections based on occupation, some differences in these incentives and controls are unearned and therefore do not meet an equity standard of fairness for distribution of rewards. However, procedural fairness can be maintained if these incentives are clearly related to the need to retain experienced personnel. We include these established mechanisms among our alternatives only because they are available to offset reducing the weight of occupational differences in promotion selections. Their costs could be funded through across-the-board decrements in other pay accounts.

Relax Lateral-Entry Rules. An externalized labor market has the potential to decrease HRM costs. Moreover, as existing nonmilitary technologies are imported into military usage at an ever-increasing rate, the services may find it necessary to externalize their military labor markets through direct appointment, at middle or upper grades, of individuals with needed technical skills. Since the services currently have the option of meeting such needs through civil service hiring or outsourcing, the case for increasing lateral entries into military service is strongest when there is a demand for bringing in *uniformed* personnel at advanced grades.

Uniformed personnel are required for positions in which incumbents must be deployable into combat and must be subject to strict military control. Successful deployment, however, depends on more than technical skills brought in from an external source. It requires general skills acquired through military training and experience. In some cases, it might be easier for experienced military personnel to acquire special technical skills than for technically qualified individuals to acquire general military skills.

Other implications are mixed. The alternative is compatible with an increasing trend toward interfirm mobility in the private sector. Inexperienced higher-grade members could undermine respect for hierarchical authority and could detract from shared military cultures by placing members who are less than fully acculturated in positions of influence. If used selectively, however, it would serve to differentiate the cultures of activities that use it, possibly by importing useful values or perspectives from other organizations.

Permitting outsiders to compete for positions of higher responsibility might be viewed as inequitable by eligible military members, especially if they believe that someone should "pay his or her dues" prior to assuming a leadership role. Our field research revealed these sentiments. Finally, transaction cost economics theory predicts that costs will be lower when entry from external labor markets is permitted under appropriate circumstances. In our survey, members generally reacted unfavorably to this alternative.

Liberalizing the movement of personnel from reserve to active-duty status could provide some of the advantages of relaxed lateral entry while avoiding some of the disadvantages. Reservists would presumably possess general military skills, allowing them to be easily absorbed at middle or upper grades in the active-duty workforce. The range of technical skills that could be acquired thus is limited, of course, to those found among reservists. Also, reservists tend to have less on-the-job experience than their active-duty counterparts. Thus, an alternative such as this might best focus on reservists with civilian jobs similar to their reserve duties.

Increase Portability of Deferred Benefits. If military labor markets are to be externalized, this alternative would be needed as a complement to increased use of lateral entries. It would be needed to make military service attractive to potential lateral entrants, who are likely to value opportunities for further lateral moves. It would also be needed to sustain fairness in advancements. If individuals outside the organization are allowed to compete for advanced placement within the organization, individuals within the organization must compete for advancement elsewhere to hold their advancement opportunities constant. However, competing outside the organization is costly to the individual member if deferred benefits are not portable.

This alternative would likely decrease retention at all points between the new and current vesting points. The resulting decrease in productivity would offset whatever was gained by attracting needed lateral entrants. It would bring military HRM closer to the early vesting in retirement benefits that is required by law for U.S. private-sector businesses. The alternative could be structured so that costs are comparable to those of the current system, although the resulting force would probably have different retention and experience characteristics.

Permit HRM Policies and Practices to Differ Across Organizations and Functional Communities. Tailoring an HRM to the specific needs of a service or functional community is likely to increase productivity; however, fairness issues must be considered. We have found in our base visits that individuals accept differentiation of rewards across groups if such differentiation is related to the arduousness of duties, particularly physical demands or frequent and lengthy deployment. Since differences in arduousness are likely to correlate more highly with functional community than with service, differentiation by functional community is likely to be seen as meeting an equity standard of fairness, while differentiation by service is likely to be seen as violating an equality standard.

This alternative implies greater flexibility in accommodating different forms of organization across services or functions and a greater capacity to develop differentiated traits and behaviors. It would, however, reduce the elements of military culture that various units hold in common. Reducing the sense of shared culture could adversely affect individuals' readiness to sacrifice their own self-interests for the good of the organization. It could also cause difficulties for individuals rotating among units (e.g., from units where individual performance is stressed and rewarded to units where teamwork is stressed and rewarded). Costs can be held budget-neutral.

Our base visits suggest that greater differentiation of rewards contingent on performance is more likely to be accepted in noncombat units than in combat units. If this occurs, it could have the intended effect of drawing people high on specific strengths to noncombat units in which those strengths might be associated with highly differentiated rewards. However, it might also have the unintended effect of drawing those same people away from combat units that need their strengths but are unable for various reasons to offer differentiated rewards.

Eliminate Differences Between Benefits for Members With/Without Dependents. Eliminating the perceived unfairness of this differentiation might have some mildly positive productivity effect, simply through eliminating the dysfunctional behaviors that accompany perceptions of unfairness. However, we note that conflicting fairness standards seem to apply.

Some see this as a housing and subsistence issue, with need as the relevant standard. Others see it as a total-compensation issue, with equity as the relevant standard. In implementing this alternative, both fairness standards can be met by elevating the allowances of all members to with-dependents rates—a costly solution. Other approaches (e.g., reducing first-termers to without-dependents rates and raising all others to with-dependents rates) would be less costly but would seem to violate a need standard. Finally, this alternative would move military HRM closer to common

U.S. practice, where differentiation of pay on the basis of dependent status is not common.

Eliminate Mobility Disincentives. If the economic penalties associated with needed deployments or moves were eliminated, improved alignment of individuals with organizational needs and greater frequency of voluntary movement to meet needs would probably increase productivity. Additionally, high mobility appears to be valued in military cultures. The existing penalties also raise several fairness issues. Because deployments and relocations are not evenly distributed, some members are penalized more heavily than others, violating an equality standard.[14] Since the services can (somewhat opportunistically) require deployment or relocation involuntarily, and because the penalties would be costly to alleviate, the penalties have persisted.

Raise Enlisted Pay Levels. This alternative would increase productivity by reducing financial management problems that distract junior enlisted members and their supervisors from their primary duties.[15] It might also enhance the productivity of senior enlisted members by reducing dysfunctional reactions to their perceived underpayment relative to junior officers. Greater compression between officer and enlisted pays would promote a more cohesive workforce, evoking the benefits of a shared culture. For senior enlisted personnel, this alternative would support an equity standard of fairness, whereas among junior enlisted personnel, it would support a need standard. The cost, of course, would be very high.

Overall Assessments

From the foregoing, we can make some generalizations about the implications of more-differentiated HRM systems:

- **Productivity.** Although all alternatives were initially identified and included in this study because they were thought to yield favorable productivity, our productivity assessments are not uniformly positive. Some alternatives affect productivity in multiple ways and perhaps in opposite directions, as in the case of individual performance rewards (performance bonuses, merit pay raises), which can increase individual productivity but reduce collaboration within a group.

- **Compatibility with U.S. Practices.** All observable effects are positive, suggesting that a military workplace with more-differentiated HRM would look much more like the private sector.

---

[14]In the case of subsistence allowances during deployments, it can be argued that the appropriate fairness standard is need. Since members are provided rations in kind when deployed, they do not need to retain an allowance provided for them to mess separately. However, for family households, where economies of scale may make the marginal cost of feeding the member less than subsistence allowance, we believe that family needs would outweigh the military member's individual subsistence needs in evaluations of fairness.

[15]Financial responsibility is a valued and expected behavior among military personnel. Commanders and supervisors are expected to provide counseling to help members keep their financial obligations in line with their resources, and to take disciplinary action against those who routinely fail to do so.

- **Military Organization and Culture.** Greater differentiation of HRM systems tends to support flexible organization structures and cultures that are differentiated across services or elements within them. The price, in some cases, is weakening of hierarchical structures and the military's culture of shared sacrifice and service.
- **Fairness.** Double standards seem to apply in many cases and almost always yield conflicting assessments.
- **Accountability.** In a few cases, assessments indicate where a lack of alternatives to a hierarchical pattern could present problems.
- **Cost.** Some alternatives have no cost implications, some have costs that can be readily offset, some entail risk premiums, and some entail major administrative or payroll costs.

Readiness for Implementation

In thinking about implementation, we divide the alternatives into two broad categories—how much they cost (low versus high) and how much statutory or policy adjustment is required (limited versus significant). Figure 3.2 shows how the alternatives may be categorized along these dimensions. Those that require limited additional legislation or adjustment of DoD policy and that have little or no effect on cost are considered available for immediate implementation. These alternatives are currently in the "toolbags" of the services that choose to use them. The remainder, which either require significant statutory or policy adjustments or have high costs, are not ready for immediate implementation. Those requiring statutory or policy adjustments would have to be tested or demonstrated in military environments, while those that appear to have potentially significant effects on cost require more-detailed cost/benefit analyses.

Low Potential Effects on Cost/Limited Statutory or Policy Adjustments Needed. The four alternatives in this quadrant in Figure 3.2 are available for immediate implementation. Three of them—reduce the weight of human capital development, reduce the weight of occupational differences, and increase special pays/ bonuses—relate to personnel management functions (promotion and retention incentives) over which the services currently have considerable latitude. The services need neither legislative relief nor changes in DoD policy to implement the human capital alternative. The Air Force, for example, recently announced a reduced weight for human capital development for its 1996 majors' board.[16]

With respect to the occupational-differences alternatives, the services follow varying practices. The Air Force enlisted promotion system is based on a principle of equal

---

[16]In a change from past practice, information on advanced academic degrees, other than those earned through service-sponsored programs, was not made available to the promotion board. The purpose of this change, according to the Air Force Chief of Staff, General Ronald R. Fogleman, was to "level the playing field," because some officers are in career fields that make it difficult to complete advanced-degree programs (Jordan, 1996).

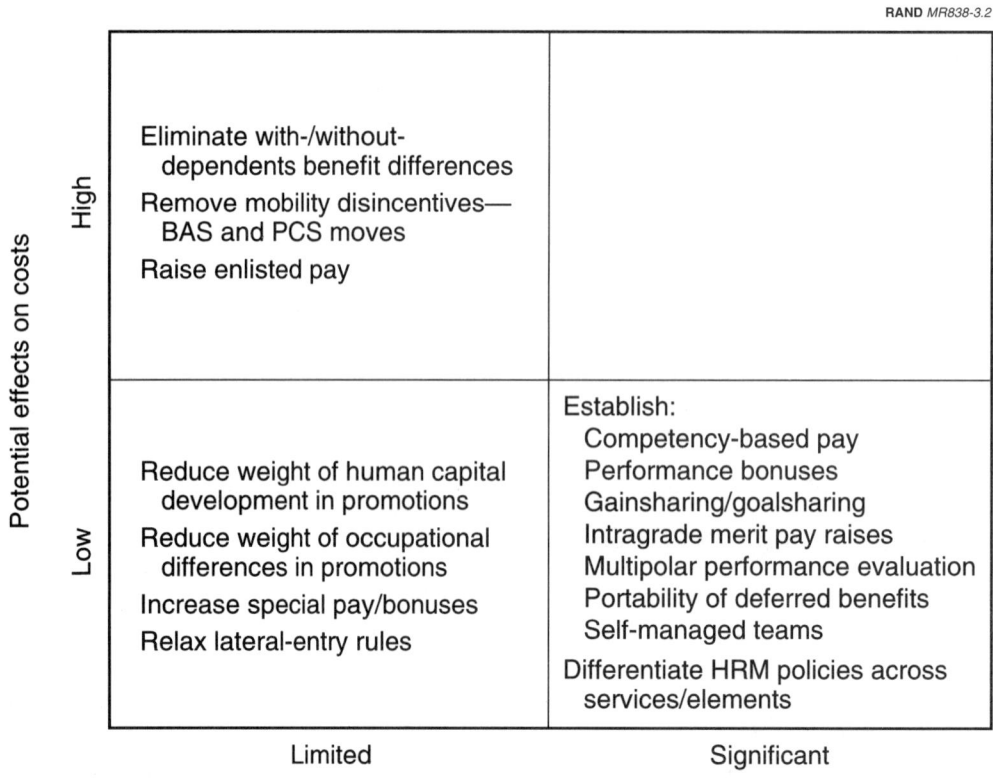

Figure 3.2—Alternatives As a Function of Potential Effects on Costs and Need for Statutory or Policy Adjustments

selection opportunity across occupations, whereas the other services allow promotion opportunity to vary widely across occupations (typically as a function of vacancies). Likewise, the existing statutory and policy framework for retention incentives would not preclude the increased use of special pays to offset a reduced weight of occupational differences in promotion processes.

The fourth alternative—relax lateral-entry rules—is currently used in a few functional communities, notably for medical professionals and conservatory-trained musicians. It could be modestly expanded under current statutes and policy (however, we encountered no other likely candidates during our field studies), but we would not advocate widespread use without corresponding adjustments in the portability of deferred benefits, which would require significant policy adjustments.

Low Potential Effects on Cost/Significant Statutory or Policy Adjustments Needed. Several of the alternatives in this quadrant in Figure 3.2 are expected to have generally favorable results but, owing to their novelty in a military setting, require deliberate approaches to implementation (as described in Chapter Four). These include several new forms of compensation—competency-based pay,[17] performance bonuses, and gainsharing/goalsharing—that require legislation and some budgetary adjustments (which, except for risk premiums, might generally be budget-neutral). To empirically evaluate their effects and to facilitate possible full-scale implementations, we recommend testing or demonstrating these alternatives on a limited scale.

Portability of deferred benefits and intragrade merit pay raises would alter military HRM systems to a greater degree than the other new forms of compensation listed above. Portability would fundamentally change the retired-pay component of compensation and has the potential to significantly shift retention patterns. Merit pay raises would fundamentally change the way basic pay is determined. Additionally, their effects will unfold over a long period of time.

Another alternative—differentiate HRM across services or functional communities—can be seen in many special pays and other current HRM practices, such as those for aviation, joint duty, and acquisition communities. However, while differentiation itself would not be new, any new form would likely require new legislation and also a willingness, reflected in new policy directions, to accept more differentiation. A service-unique basic pay table, for example, would represent a significant departure from past practice.

Multipolar performance appraisals and self-managed teams require shifts in service policies about performance evaluation. Since no statute or DoD policy governs military performance evaluations, the services currently have the latitude to implement variants of these alternatives. However, they represent significant departures from current service policies, which are predicated on strictly hierarchical forms of accountability.

The only formal applications of multipolar performance evaluations that were brought to our attention by focus-group participants were in pre- or immediate post-commissioning programs. From the generally negative experiences individuals reported with those applications (lack of objectivity by peers in a competitive environment), we would advocate careful testing of expanded use of this approach and recommend introducing it as a developmental tool prior to using it as a basis for promotion or performance awards.

High Potential Effects on Cost/Limited Statutory or Policy Adjustments Needed. The three alternatives placed in this quadrant—eliminate the differences between benefits for members with/without dependents, remove mobility disincentives (BAS and PCS moves), and raise enlisted pay levels—entail costs that would likely be very high, with no apparent first-order offsets (as we saw for some of the other pay-related alternatives).[18] We recognize that some of these alternatives could yield productivity

---

[17]Competency-based pay is currently authorized by statute in some forms. For example, special pay is authorized for proficiency in foreign languages. However, new forms would require additional legislation.

[18]With regard to the "remove with-/without-dependents benefit differences" alternative, the Seventh QRMC (1992, MTS 1, p. C-22) estimated a cost of about $3 billion to raise the pay of all members to with-dependents rates. That analysis also discusses less-expensive alternatives (reducing all pay to without-dependents rates or a combination of without-dependents rates for first-termers and with-dependents rates for the career force).

gains that, when netted against costs, could produce second-order savings. However, analysis to support such a conclusion is beyond the scope of this study. Detailed cost/benefit analyses would be needed to support adoption of any of these alternatives.

Chapter Four

IMPLEMENTATION ISSUES

In the preceding chapter, we divided the 16 alternatives into three categories (shown in Figure 3.2) as a way of thinking about their readiness for implementation (independent of whether DoD would decide to implement them). The alternatives that have low potential cost and limited requirements for policy or statutory shifts (the lower-left quadrant) are all performance-related changes in promotion practices that are currently within the purview of the military services. Dissemination of the Eighth QRMC's research findings and publication of this document by RAND should help these ideas to enter and perhaps win acceptance in the services' policy deliberations. While we believe that movement toward these alternatives would help the services to shape performance, the services are in the best position to decide the appropriate weighting of performance and other objectives within their promotion systems.

Another category includes alternatives that have potentially high costs but that are not innovations (the upper-left quadrant). Because of their high costs, they warrant more-detailed cost/benefit analysis, beyond the scope of this research, prior to testing or implementation.

Here, we are primarily concerned with the alternatives we identified in the third category (the lower-right quadrant), which require significant statutory or policy shifts.

In this chapter, we first discuss what we learned from private-sector firms we visited about the movement toward more-differentiated HRM systems. Following that, we describe specific steps leading to implementation of those alternatives requiring significant statutory or policy shifts. For these cases, we argue that a phased approach to implementation seems warranted. Since a phased approach appears to be useful in introducing such innovations, we then discuss demonstration projects—a public-sector avenue for the phased approach.

INSIGHTS FROM PRIVATE- AND PUBLIC-SECTOR IMPLEMENTATIONS

The same forces that may be encouraging the military to rethink the role of the HRM function—i.e., looming resource scarcity and growing environmental uncertainty—have long been present in the private sector, as well as in other parts of the public sector. Increased competitive pressures, deregulation, privatization, and changing patterns of consumer demand have all forced many public- and private-sector organizations to significantly redesign both their organizational structures and supporting HRM policies.

As part of its research support for the Eighth QRMC, the RAND team wanted to incorporate lessons learned from other ongoing organizational restructuring efforts. For this reason, the team selected half a dozen public- and private-sector organizations for in-depth research and analysis. The sample of organizations we visited include a state government, a public utilities/telecommunications company, two defense electronics firms, an automobile producer, and an office equipment/document company.

All six of those organizations were shifting away from bureaucratic and hierarchical organizational structures toward higher levels of flexibility, decentralization, and cross-functional collaboration. Decisionmaking authority and accountability for business results were being shifted to lower levels in each organization. Production and service delivery were being reorganized along cross-functional lines for better alignment with important business processes. The more-innovative organizations we visited were also moving toward making process improvement a regular part of their company operations rather than an extraordinary event. These companies had put systems in place to support both organizational learning and continuous improvement.

To support the emphasis on increased decentralization and flexibility, the six organizations were also moving toward a more strategic approach to HRM. Day-to-day decisionmaking on human resource policies was being shifted to program and division managers, and many of the routine administrative procedures typically handled by HRM personnel were being automated. Increasingly, the main responsibility of the HRM department in these organizations was to create a strategic HRM framework that changed both incentives and organizational culture in ways consistent with emerging business needs.

Shifting from a bureaucratic and hierarchical organization to one that places a premium on flexibility and innovation is difficult and complicated. Successfully implementing change requires careful planning and systematic inclusion of key stakeholders. The primary elements for a successful implementation strategy in the six organizations we visited were leadership by senior executives in setting a strategic direction, participation by middle managers and employees in redesigning processes, a phased approach to implementing new policies, and a deliberate communication effort.

Strategic intent, typically provided by senior management, can be defined as a clearly articulated sense of purpose that provides the emotional and intellectual energy for accomplishing an organization's goals (Hamel and Prahalad, 1994). Strategic intent becomes especially important when an organization moves away from bureaucratic controls. Decentralizing authority and empowering workers does little good if employees lack a clear and compelling sense of direction to guide them.

Having developed a strategic vision, senior management alone does not have the breadth of expertise nor the immediate proximity to the organization's day-to-day operations to develop a course of action. The majority of organizations in our study involved second-tier and other managers in laying out a blueprint for organizational change. Participation from organizational representatives not only developed a greater sense of commitment to the idea of change but also provided clarity about what the change should look like.

The blueprint established a framework that allowed organizations to change individual processes in a coordinated fashion. It provided a tangible mechanism for ensuring alignment among core-process changes and among the HRM systems responsible for developing and encouraging required competencies. Having identified potential improvements and the support systems required to promote their adoption, most of the organizations we visited chose to implement their change efforts in an evolutionary, as opposed to revolutionary, fashion. Most organizations pilot-tested new approaches in individual business units and then phased them in gradually across the rest of the organization. This phased approach enabled corporate officials to gain experience with the new system, identify problems, determine solutions, and develop effective communication strategies.

One of the most consistent issues conveyed to us during our interviews was the importance of an effective communication strategy. Several key communication lessons emerged from our interviews. First, most companies agreed that it was important to start planning communication early in the change process. Second, while several organizations recognized that they had not communicated effectively in the past, most agreed that communication could be vastly improved by developing an integrated communication strategy up front. Finally, all the organizations in our sample cited supervisors and middle management as the single most important factor for ensuring effective communication. Because most employees look to their supervisors to interpret organizational events, it is important that supervisors and managers know about, understand, and be committed to change.

IMPLEMENTATION STRATEGIES—THE NEED FOR A PHASED APPROACH

Taking a lead from the implementation practices of the innovative organizations we visited, we suggest that an appropriate role for managers of the central HRM function in DoD is to prepare the way for evolutionary adoption of strategic HRM choices by *business units*—services, functional communities, or other subordinate elements—within the Department.[1] DoD, in this view, is too diverse to be well served by a single encompassing strategic vision and an undifferentiated HRM system. Alternatively, given a number of considerations ranging from inertia to interoperability to battlefield exigencies, DoD will likely find grounds for retaining some common structures and processes across its business units. Thus, we anticipate that the most likely path for differentiation of military HRM is tailored use of existing flexibilities,

---

[1]Our use of the term "business unit" conforms to a convention adopted by the Eighth QRMC to denote any element within DoD that has a mission or vision sufficiently distinctive to warrant a strategically differentiated HRM system.

such as the latitude given the services to fashion their own promotion systems, plus selective adoption of innovations by business units to meet their strategic needs.

That said, the implementation strategy we describe here is intended to be that of the DoD central HRM system manager rather than that of the business-unit manager. Implementation from the perspective of a business-unit manager is an exercise in tailoring the HRM system to specific business-unit needs, while implementation from the perspective of a central HRM system manager is an exercise in making appropriate tools available to business units. Although implementation at the business-unit level is the ultimate objective of a strategic approach to HRM, it is a case-by-case process that is beyond the scope of this research.

When we examine the alternatives in the "requires significant policy or statutory shifts" category, we believe that new compensation-related incentives for performance and human capital development should be introduced using the phased approach that private-sector firms seem to favor for their HRM innovations. In a *phased approach*, central HRM system managers would first assess a need for new or enhanced tools, either in specific business units or across many business units, as we have sought to do in this research. Alternatives would be developed and evaluated, ideally with members of relevant workforces involved in the process. The most-promising alternatives would be pilot-tested on a limited scale. Given a successful test, the innovation would gradually spread as business managers, at their option, found it a useful one to adopt.

A phased approach reduces risk and facilitates change in several important ways. First, the consequences of these alternatives in a military environment are not fully known, and, thus, testing on a limited scale is prudent. Second, the test population provides a source of worker participants interested in designing and evaluating alternatives. Third, confidence in the effectiveness of new alternatives and acceptance of the changes they require can be cultivated by allowing stakeholders to observe concrete successes. Fourth, selective adoption by business-unit managers naturally affords differentiation across business units.

Finally, legislation is needed to introduce these changes in military compensation.[2] Legislation permitting limited testing (or *demonstration projects*, as they are commonly referred to in federal legislation) would likely be easier to enact than substantive statutory changes. The demonstration-project approach provides agency flexibility without sacrificing political oversight. The successful conduct of numerous demonstration projects in civil service personnel management, under the provisions of 5 USC 4703, serves as a ready model.

Multipolar performance appraisals and self-managed teams are like performance and human capital pay incentives in that they lend themselves to a phased approach, most likely as part of a bundle of other changes geared to the needs of a specific

---

[2] 5 USC 5536 states that "[a]n employee or a member of a uniformed service whose pay or allowance is fixed by statute or regulation may not receive additional pay or allowance . . . unless specifically authorized by law. . . ."

business unit. They are unlike new pay incentives in that they do not require legislation or any action by DoD central HRM system managers.

Portability of deferred benefits and intragrade merit pay raises, as mentioned in the preceding chapter, require fundamental structural changes and have long-term effects. Thus, testing on a limited scale would be difficult to arrange and possibly inconclusive. As a result, simulation modeling might provide a better approach to policy analysis for these alternatives.

In implementing or demonstrating new approaches to military HRM, attention should be given to corresponding approaches for civil service employees. All else being equal, consistency of approaches for military and civilian personnel working in the same functions would probably reduce perceived inequities associated with inconsistent treatments. Where military members and civilian employees work closely together in tasks with low separability, consistency would be especially important. Ideally, future military and civil service demonstration projects within DoD would be integrated.

DEMONSTRATION PROJECTS

Demonstration projects offer DoD central HRM system managers a convenient means of developing and introducing new tools. However, they serve multiple and, possibly, conflicting purposes. On the one hand, as part of a broader implementation strategy, demonstration projects serve as instruments to introduce change. They provide avenues for demonstrating the effectiveness of a new approach to perhaps skeptical stakeholders, for co-opting managers and employees by involving them in the design of the project, and for fine-tuning the administrative details of a new approach. Properly constructed for this purpose, they help "sell" a new idea and contribute to a smoother implementation.

On the other hand, demonstration projects serve as scientific experiments. They permit policymakers to evaluate competing strategies, analyze variations, or examine consequences in varying environments. Properly designed, they provide an empirical basis for ongoing policy analysis.

These purposes will inevitably conflict in some respects. Nonetheless, both purposes are important. An alternative is most likely to become a useful tool if some champion with a stake in its success deliberately pursues implementation. But legislators and central HRM system managers will look for something stronger than enthusiasm as a basis for necessary statutory or policy adjustments. Some compromises are necessary if a demonstration project is to serve both purposes.

Conditions for Successful Projects

In view of their multiple purposes, demonstration projects require a careful balance of enthusiasm and objectivity. On the one hand, enthusiasm for a significant innovation on the part of individuals interested and involved in it is probably necessary to overcome organizational inertia. On the other hand, overenthusiastic support for a demonstration concept can contribute to a *Hawthorne effect*, whereby individuals behave differently because of the attention they receive as participants in an experiment rather than because of the experimental variables; can restrict the range of options tested; or can cloud objective evaluation.

The right combination, we believe, is an enthusiastic champion of the concept within the implementing organization and an impartial evaluator outside the implementing organization. For civil service demonstration projects, executive agencies propose projects and play the role of enthusiastic champions while the U.S. Office of Personnel Management (OPM) oversees the evaluation. For military personnel demonstration projects, it appears appropriate for services, defense agencies, or functional communities to propose and champion projects while OSD oversees the evaluation.

To ensure that the demonstration concept has sufficient resources and that responsible commanders and managers are given sufficient scope, policymakers and high-level managers should agree widely that the demonstration project is worthwhile. Implementation at one level of policy made at another level may require multiple decision points and clearances, and apparently simple sequences of events may actually depend on complex chains of reciprocal interaction (Pressman and Wildavsky, 1973).

Successful implementation may depend on cultivating the needed buy-ins. Thus, at the OSD level, we recommend obtaining the concurrence of at least the Secretary of Defense, the Under Secretary for Personnel and Readiness, and the Assistant Secretary for Force Management Policy, and any staff elements with an interest in the local activities involved in the demonstration. At the service level, concurrences should be sought from the service Secretary, Chief of Staff or equivalent, secretariat and uniformed HRM chiefs, and functional managers responsible for the activities affected by the demonstration project. At major command or other intermediate headquarters levels, project leaders should consult with the commander, chief of the HRM function, and affected functional managers. At local levels, they should consult with managers of the affected activities, HRM managers, and commanders with scope over them.

Functions and organizations selected for demonstrations should be those whose environments are, ex ante, judged to be appropriate for the demonstration concept. In a purely scientific experiment, it might be useful to test a concept in a wide range of environments, providing a richer data set with which to examine differences in effect. However, in experiments with HRM systems, a failed approach might well have negative consequences for people subject to it and also might jeopardize the momentum for change that demonstration projects are intended to create or sustain. Thus, to minimize risks to the individuals involved and to avoid derailing otherwise useful approaches, demonstration projects are best undertaken in environments in which the risks of failure are low.[3] In analytic terms, these considerations introduce

---

[3]We do not intend to imply that demonstrations should be avoided in circumstances where meeting individual or group performance goals is risky or that alternatives involving *pay at risk* should not be

several disadvantages. A restriction-of-range problem limits robustness of potential findings. A lack of randomness makes it difficult to generalize findings to other cases. In the interest of avoiding derailment, implementation might have to proceed incrementally, with earlier demonstrations in more-propitious environments and later demonstrations in less-propitious ones.

Another approach is to test a new alternative simultaneously in multiple environments that are stratified, to reveal the strengths and weaknesses of the alternative. This approach would provide a more robust test of the alternative and would permit the alternative to be propagated more rapidly throughout the range of environments in which it would be found useful. A risk of derailment arises from the potential harm to, and resulting negative attitudes of, individuals in environments in which the alternative had little chance of success. For example, testing a pay-at-risk scheme in a cross section of military units would be unfair to individuals in units for which performance metrics have not been successfully developed. Moreover, since legislative relief to permit demonstration projects has typically placed tight constraints on the number of organizations and individuals involved, the number of units of analysis required for a more robust experimental design may not be available.

A fundamental step in designing a demonstration project is identifying the outcomes the demonstrated approach is expected to improve. Outcomes of interest would generally include some broad measure of organizational performance, but might also include individual performance or instrumental attitudes and behaviors that contribute to better performance. Functional and local activity managers should be consulted early in the design of the project to ensure a thorough understanding of desired outcomes and to gain insight into how to measure them. Agreement on desired outcomes is important among those who must buy into the demonstration project and useful among other business units that might later be expected to adopt the demonstrated approach. The methodology described in Appendix A could also be used in a focused way to explore the outcomes and behaviors of interest within a business unit.

The design of demonstration projects should be sensitive to analytical needs. Demonstration projects are ideally conducted in activities that can be matched to appropriate control groups for cross-sectional analysis. To permit time-series analysis, historical files containing appropriate measures of organizational effectiveness and worker attitudes must be located, or a baselining period, before the demonstrated approach is introduced, must be set aside to collect relevant measures.

In many projects, outcomes of interest (e.g., productivity) may be measurable only at organizational rather than individual levels. Thus, units of analysis are likely to be organizations and are likely to be few in number. Consequently, experimental and control sample sizes may be too small to rely on randomness as a control for other variables. Thus, to the extent possible, measurement must not be limited to the in-

---

tested. Rather, we believe such alternatives should be tested only when conditions are suitable for them. *Suitable conditions* might include, for example, the availability of measurable outcomes that are highly correlated with individual or group effort/performance.

dependent and dependent variables of interest. Researchers must identify other environmental variables affecting important outcomes. These other variables must be measured and their influence must be isolated through regression analysis or other appropriate statistical techniques.

Legislative Considerations

Many promising new alternatives require changes in legislation to give policymakers sufficient flexibility. In particular, changes in compensation systems require relief from the provisions of 5 USC 5536, which states that any federal employee or member of a uniformed service may not receive any pay or allowance that is not specifically authorized by law. For civil service employees, 5 USC 4703 provides blanket relief from this and other statutory restrictions in demonstration projects designed to test HRM alternatives. No similar relief exists to allow demonstration of HRM alternatives for military personnel.

The statutory environment complicates the task of policymakers contemplating a phased approach to differentiating military HRM systems. Figure 4.1 illustrates three alternative approaches to dealing with this complication. One alternative, labeled "limited approach" in the figure, is to restrict strategic HRM planning to the range of alternatives available under current law, which would block many potential enhancements. Another possibility, labeled "'trailblazer' approach," is to progress through a strategic-planning process with a well-disposed business unit up to the point at which statutory limits are encountered, then seek either specific legislative relief or general relief in the form of a military HRM demonstration statute parallel to the civil service statute. This approach would place a lengthy wait for enabling legislation on the critical path of a strategic planning and implementation effort, with possibly irrecoverable loss of momentum. Additionally, if the legislative relief is specific to the case at hand, it would necessitate making additional time-consuming loops back through the legislative process in the event that pilot testing reveals needed modifications. A third, "proactive" alternative is to seek broad legislative authority for demonstrations, anticipating its probable need in almost any strategic adaptation of the HRM system.

Legislative relief could parallel, with minor changes, the civil service demonstration project authority provided in 5 USC 4703. Changes might include substituting DoD for OPM as the authority for conducting and evaluating demonstrations, removing references to labor agreements, and making minor adjustments to align the language more precisely with military law and regulation.

Specific Cases

We anticipate both Congress and internal DoD sources will exert strong pressures to conduct a demonstration project involving incentives for members of the acquisition workforce. Well-publicized inefficiencies in an expensive defense acquisition system have created a focus of interest, resulting in a continuing series of reforms aimed at acquisition processes and the workforce that manages them.

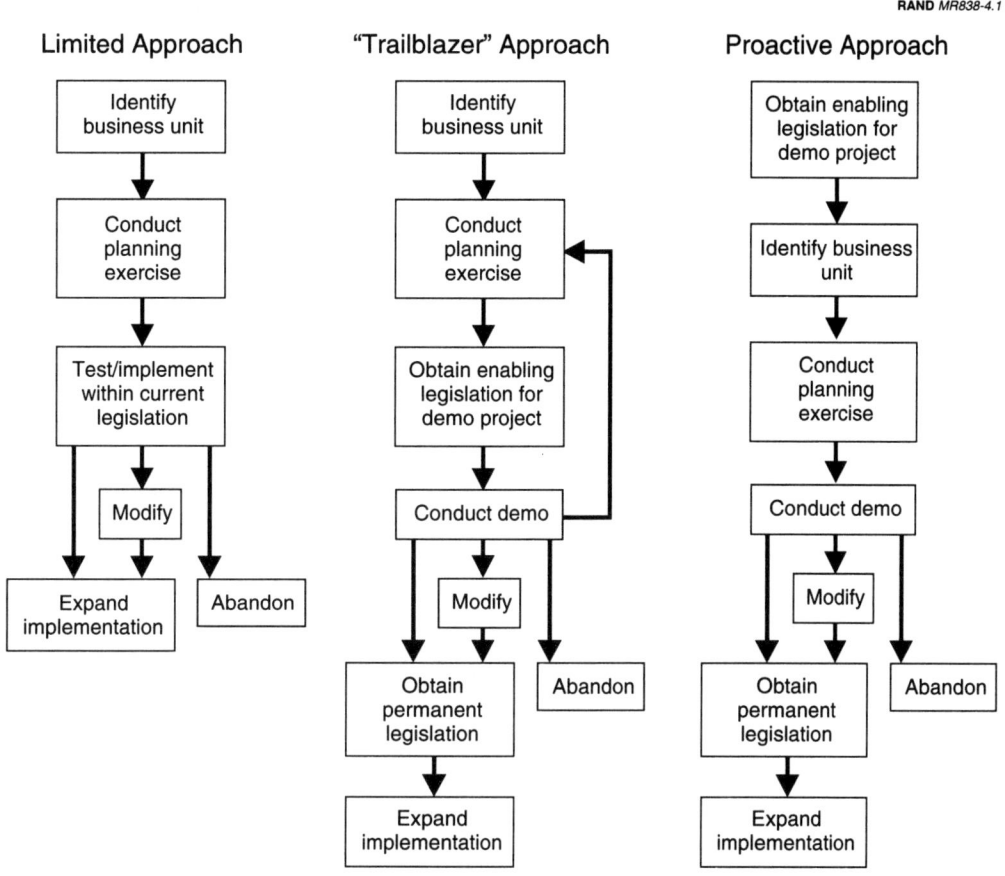

Figure 4.1—Alternative Approaches to Accommodate Legislative Constraints

In 1990, Congress passed the Defense Acquisition Workforce Improvement Act (DAWIA),[4] which established training, education, and experience prerequisites for individuals performing critical acquisition duties. Additionally, the Federal Acquisition Streamlining Act of 1995 (Public Law 103-355) required that the Secretary of Defense provide an enhanced system of incentives, including pay for performance, *to the maximum extent consistent with applicable law*, to encourage excellence in the management of defense acquisition programs (Sec. 5003[b]) and to recommend any additional legislation needed (Sec. 5003[c]).[5] More recently, in the Federal Acquisition Reform Act of 1996 (enacted as part of the 1996 Defense Authorization Bill), Congress encouraged the Secretary of Defense to commence a demonstration project to "determine the feasibility of . . . improving the personnel management policies or procedures with respect to the acquisition workforce of the

---

[4]DAWIA was enacted as part of Public Law 101-510. It is now incorporated in 10 USC 1701 et seq.

[5]As of this writing, no initiatives or legislative proposals have been forwarded from the Department of Defense.

Department of Defense."[6] In a section of the Act that applies to nondefense agencies, Congress specified performance incentives for acquisition workforces that:

> (A) relate pay to performance (including the extent to which the performance of personnel in such workforce contributes to achieving . . . cost goals, schedule goals, and performance goals . . .)
>
> (B) provide for consideration, in personnel evaluations and promotion decisions, of the extent to which the performance of personnel in such workforce contributes to achieving such cost goals, schedule goals, and performance goals.[7]

DoD and the services have shown similar enthusiasm for streamlining their acquisition processes and modifying HRM systems to enhance the performance of their acquisition workforces.

We see several advantages and disadvantages in tying the pressures for acquisition-workforce reform to the need for military HRM demonstration authority.

An advantage is that the apparent readiness of Congress to provide the needed legislative relief for the acquisition community could clear the way for other demonstrations in less-visible environments. Arguably, DAWIA certification requirements have already differentiated the HRM of military acquisition professionals to some degree, setting the stage for further strategic adaptations.

One disadvantage is that defense acquisition programs may not lend themselves to outcome-based accountability, of which pay for performance is a variant. Other disadvantages are that multiple principals frequently establish conflicting goals and objectives; programs frequently require trade-offs among performance, cost, and schedule that cannot be anticipated a priori; and projects are "lumpy," meaning they can be evaluated only after completing multi-year cycles (Mayer and Khademian, 1996). Before conducting a performance-related HRM demonstration project in the acquisition community, HR and functional managers should determine if intermediate and shorter-term outcomes can be identified. Another concern, given the history of acquisition-specific legislation, is that demonstration authority might also be acquisition-specific, which, while it might contribute to enhancing acquisition management, would limit the capacity of HR managers to test similar improvements in other environments.

Broad statutory relief to permit demonstration projects for HRM innovations might well require a specific "trailblazer" application to help propel it through the legislative process. If that proves to be true, we recommend finding several possibilities outside of or in addition to the acquisition community and where legislative relief is less likely to be case-specific and line of sight from behavior to outcomes is more likely to be clearer.

The business-unit setting for a trailblazer project would have two characteristics: (1) have at least some requirements for uniformed military personnel, so that complete

---

[6]Public Law 104-106, para 4308, February 10, 1996.
[7]Public Law 104-106, para 4307, February 10, 1996.

civilianization or outsourcing is not an option (obviating the need for a military HRM system); and (2) not be directly combat-related, because new HRM flexibilities tend to be less useful in combat units. Logistics, training, professional (medical, legal, chaplain), or recruiting communities might offer possibilities.

Chapter Five
CONCLUSIONS

Military HRM appears to be much less differentiated than the systems of many high-performance private-sector firms. Nonetheless, differentiation is effective for such important behaviors as motivation of performance, retention, human capital formation, and others. While private-sector firms have the advantage of greater flexibility and higher ceilings for extrinsic rewards, the military services enjoy the advantages of powerful intrinsic rewards. Expending effort to provide national defense apparently requires less economic incentive than expending effort to further stockholder interests.

Although existing systems work well, we see opportunities to prudently increase differentiation in military HRM. Increased differentiation might occur at an individual level, if explicit forms of pay were used to reward performance, human capital development, or other behaviors. It might also occur at a structural level, if an incentive pay were provided in one functional community but not another or if the form of the incentive differed across units, services, or functional communities. In either case, increased differentiation must be approached cautiously because of strong and sometimes-conflicting standards of fairness applied by military personnel in evaluating their HRM systems. Additionally, differentiation of extrinsic rewards must not be allowed to undermine highly effective intrinsic rewards and other important features of military culture.

Future environments may demand more differentiation, and these demands may vary by functional community. For both reasons, a flexible HRM system that can respond in a timely and effective manner to varying needs may be desirable. Since sensitivity to changing environments is likely to be greater at decentralized levels of defense organizations, responsiveness would seem to be supported by an outward and downward movement of authority for tailoring HRM systems. In broad terms, we foresee core combat units continuing to successfully employ traditional and hierarchical forms of HRM, with innovation proving more successful in non–core combat activities.

We identified a wide range of potentially useful innovations or enhancements. However, they vary considerably in their readiness for implementation: Some are currently in the toolbags of the services, some would benefit from a phased implementation (including careful testing), and some require further cost/benefit or other analysis. For innovations that would benefit from phased implementations, demonstration projects provide a suitable avenue. Because of widespread interest in enhancing acquisition performance, HRM for the acquisition workforce is likely to emerge as a top candidate for a trailblazing application of the process. However, because of difficulties in assessing accountability and in view of other potential pitfalls related to the acquisition workforce, we see several possibilities of derailment if the acquisition workforce is used as a trailblazer; accordingly, we recommend a search for other candidates as well.

Appendix A
MILITARY BASE VISITS: METHODOLOGY AND PILOT DATA SET

INTRODUCTION

As part of our research, we conducted visits to military bases. Because of limited time and money, we decided to select just four bases, one from each service. We made our selections according to four criteria. In order of importance, the criteria were (1) representativeness, (2) completeness (including as fully as possible the range of environments experienced by members of the service), (3) diversity (providing good contrasts between environments), and (4) closure not imminent. Given these four criteria and the willingness of the base to accommodate our visit, we selected Fort Hood, TX; Norfolk Naval Station, VA; Mt. Home Air Force Base, ID; and Camp Lejeune, NC.

In conducting the base-visit research, we had two intents: (1) to demonstrate a methodology for collecting some of the data we believe future HRM designers would need to make informed decisions, and (2) to collect a pilot data set, using the methodology in (1) to demonstrate its potential usefulness for making difficult HRM design decisions.

In this appendix, we first explain the methodology we designed and discuss the forms used to gather information (both quantitative and qualitative). Then we show the raw pilot data set derived from applying the methodology. Where appropriate, data (both quantitative and qualitative) have been incorporated into the discussion in this report.

METHODOLOGY

In doing the base-visit research, we conducted focus groups, surveyed focus-group participants using two questionnaires, conducted interviews, and toured facilities at the bases listed above.[1] Below, we discuss the focus-group, survey, and interview process.

---

[1] Because the Coast Guard uses the same compensation system as the military, we also spent a day visiting Coast Guard activities in Portsmouth, Virginia, to learn about the Coast Guard's attitude toward HRM. While there, we conducted two focus groups with a total of 12 Coast Guard personnel. Because of the small number of participants, we decided not to include these data in any of the analyses that follow. The only difference we noticed between the Coast Guard and the other services was a greater interest in using

Focus Groups

We conducted eight focus groups per base (except for the Air Force, where we conducted six). Each focus group consisted of 4–12 military personnel[2] and 1–3 focus-group facilitators, and lasted about 2 hours. Within each focus group, participants were homogeneous in terms of (1) service, (2) enlisted or officer, (3) junior or senior (less than or greater than 10 years of service, YOS), and (4) occupational community (combat or noncombat). This design allowed us to sample from most of the major communities in the military, which was our best sampling strategy, given that time and money prevented us from obtaining a representative sample of military personnel.

Within each service, we examined a core combat and noncombat community. Table A.1 provides a summary of those communities. We selected combat communities that were representative of the services and available at the bases we visited. In the Army, we talked with personnel within an armor division. In the Navy, we talked with personnel from nuclear submarines and surface combatants. In the Air Force, we talked with pilots and other aircrew from a composite (fighter and refueling) wing. In the Marines, we talked with personnel from infantry units.

The noncombat functional communities varied from service to service, because we believed that the service differences within a functional community would be less than the differences across these communities. In the Army, we looked at C^4I (command, control, communications, computers, and intelligence). In the Navy, we looked at the professional communities (medical, legal, and chaplain). In the Air Force, we looked at the logistics (maintenance, supply, and transportation) community. In the Marines, we looked at the administrative (personnel, finance, etc.) community.

Military personnel within each of the eight communities above were junior (less than 10 YOS) or senior (10 or more YOS) enlisted members or junior or senior officers. We separated these four groups because we reasoned that they would have distinct thoughts and feelings and that mixing them would discourage open discussion.

Table A.1

Combat and Noncombat Communities, by Service

Service	Combat Community	Noncombat Community
Army	Armor	C^4I
Navy	Surface and Submarine Warfare	Professional
Air Force	Aircrew	Logistics
Marines	Infantry	Administrative

---

innovative private-sector HRM techniques. However, because of the small size of the Coast Guard sample, this apparent difference should be viewed circumspectly.

[2]We requested 12 participants in each group. We did not get this number in every case, owing to no-shows or nonavailability of personnel in units tasked to provide participants. Mean size of the groups was 9.9 participants.

All told, we conducted 30 focus groups involving 298 military personnel. The reason for not conducting 32 focus groups is that the Air Force core combat community consists almost exclusively of officers.

Our goals for the focus groups were to (1) develop trust and rapport with the participants so that our other goals could be achieved, (2) use general questions to get people talking about their perceptions on some HRM topics, and (3) use probes to explore their perceptions in depth. Exhibit A.1 at the end of this appendix reproduces the introductory remarks for the focus groups, and Exhibit A.2 reproduces the protocol for the focus groups.

Surveys

During the first 15 minutes and last 15 minutes of each focus group, participants completed two short, standardized questionnaires on the topics of the focus group. One purpose of the questionnaires was to collect the same critical data from everyone in an efficient way. The first questionnaire was open-ended, which gave participants the opportunity to tell us their thoughts before being influenced by our prompts or the remarks of other participants. In addition, it helped to familiarize participants with the topics of the focus group, thereby making the focus groups more effective. Finally, it served as a convenient segue for focus-group questions.

The second questionnaire was close-ended, which provided us with quantifiable data after the participants had been engaged in the topics for more than an hour. Moreover, it allowed focus-group participants to express their differences, which they may not have had a chance to do during the focus group. Exhibits A.3 and A.4 reproduce the complete text of the two surveys.

Interviews

We interviewed military personnel in charge of key functional units (e.g., squadron or battalion commanders at the O-4, O-5, or O-6 level) to learn their unique perspectives. Specifically, we asked whether they had the right tools (e.g., rewards and punishments) to motivate their troops, what their attitude was about using private-sector HRM practices in their unit, and what changes in HRM they would like to see in their units. Additionally, the commanders we interviewed gave us tours of their communities. The goal here was to learn about the diversity of tasks being done in and by these communities and the conditions (organizational, physical, economic, social) under which members of those communities work. For example, we toured a Navy submarine to learn about the living and working conditions of this unique community.

Strengths and Weaknesses of the Approach

The approach we used has two key strengths. First, although our subjects at the bases are not experts in HRM (and so cannot be expected to tell us what their HRM system should look like), they are experts in many other matters, some of which are relevant to designing an appropriate HRM system. For example, our subjects know best what motivates them, which obviously must be considered for an HRM system to be effective. Second, the approach uses several complementary methods. While each method has its own strengths and weaknesses, using several methods together, rather than any one alone, eliminates many of the weaknesses. For example, focus groups provide rich qualitative data but fail to provide easily quantifiable data. But the closed-ended questionnaire compensates for the focus groups' weakness by providing standard quantifiable data from every subject.

As with all approaches, ours has several important limitations. First, the data set we created does not provide all the information needed to make wise HRM design decisions. The most obvious omission is economic data, such as comparative military and civilian income information, to test the validity of subjects' perceptions. Second, our data set fails to resolve all the ambiguities in the complex questions we address. And third, our sample is relatively small; thus, it is not completely representative of the communities we examined and misses some important communities.

Fortunately, these limitations can be overcome. The first limitation can be overcome by collecting data on other topics vital to making informed HRM design decisions. One obvious area in need of hard data is the economic costs and benefits of HRM design options. It can also be overcome by collecting data on the same topics using subjects who have perspectives different from those of current military personnel. For example, it would be useful to ask military experts within Congress, academia, and other institutions what they think the military's future desired ends should be. The second limitation, ambiguity, can be overcome by asking more questions on each topic. Each question would need to be designed to address a different facet of each topic. The set of questions could then begin to resolve much of the ambiguity. The last limitation requires a larger, more-representative sample of all the major military communities.

Pilot Data Set

During the base visits, we collected data on the following topics:

- Behaviors that are desirable and undesirable in each subcommunity
- Behaviors that are rewarded, not rewarded, and punished (and how) in each subcommunity
- Rewards and punishments that are effective and ineffective in each subcommunity
- What is perceived to be fair and unfair in military compensation in each subcommunity
- The perceived advantages and disadvantages of importing private-sector HRM techniques into each subcommunity.

Here, we provide the results of these topics across subcommunities. Later, we examine the results for these topics by subcommunity.

Desirability and Undesirability of Behaviors

Military personnel were asked to rate the desirability of and degree of reward given for 26 behaviors. They were asked to rate desirability on a scale from 1 ("extremely undesirable") to 7 ("extremely desirable"), as their superiors would rate it. They were asked to rate how much the military rewarded each behavior on a scale from 1 ("punished severely") to 7 ("rewarded a lot"). To assess the gap between the desirability and rewardedness of a certain behavior, we created a new variable, *difference*, which is simply the difference between how desirable and how rewarded each behavior is. A negative difference score suggests that a behavior is "underrewarded" given its desirability, whereas a positive difference score suggests that a behavior is "overrewarded" given its desirability.

To analyze these data, we conducted a two-factor repeated-measures analysis of variance (ANOVA), with behavior as the first factor and difference as the second factor. The results indicate two main effects and an interaction. The behavior main effect, $F(25,7300) = 213.14$, $p<.01$, indicates that the average desirability and rewardedness of the 26 behaviors vary. The difference main effect, $F(1,292) = 905.37$, $p<.01$, indicates that, on average, behaviors are more desirable than they are rewarded. The interaction effect, $F(25,7300) = 95.15$, $p<.01$, indicates that the extent of difference depends on the behavior (i.e., the allocation of rewards is not completely sensitive to the desirability of behaviors).

Table A.2 show the results.

The Effectiveness of Rewards and Punishments

In this section of the survey, we wanted to learn what makes military personnel enjoy their work and want to work hard and well. We designed two complementary rating tasks for this purpose. In the first task, we gave focus-group participants categories of rewards and punishments (e.g., your pay) to consider retrospectively. Specifically, they were asked, "Based on your experience in the military, how much has each of the following factors influenced how hard and how well you have worked?" They rated answers on a 4-point scale (from 1 = "had no effect on your work" to 4 = "had a large effect on your work").

The results of the first task are reported in Table A.3.

In the second task, we gave focus groups definite levels of rewards and punishments (e.g., receiving $100/month more in basic pay) to consider prospectively. Specifically, they were asked to "rate how each reward or punishment would affect the quality and quantity of your work" on a 7-point scale (with 1 = "much worse," 4 = "no effect," and 7 = "much better").

The results of this task are found in Table A.4.

Table A.2

Mean Desirability and Rewardedness of Selected Behaviors

Behavior	How Desirable?	How Rewarded?	Difference
Obeying orders of my superiors	6.75	4.66	−2.09
Helping my unit to achieve its goals	6.72	5.43	−1.29
Respecting authority	6.66	4.57	−2.09
Taking initiative; acting like a leader	6.54	5.50	−1.04
Having a good attitude at work	6.38	4.80	−1.58
Working cooperatively with my peers	6.26	4.68	−1.58
Conforming to the norms and values of the military	6.25	4.62	−1.63
Being more productive at my job	6.24	5.02	−1.22
Exerting more effort at my job	6.13	4.86	−1.27
Making fewer mistakes at my job	6.11	4.69	−1.42
Maintaining physical fitness	6.08	4.77	−1.31
Thinking creatively (finding new solutions to old problems)	6.07	5.23	−0.84
Helping my peers to achieve their goals	6.05	4.86	−1.19
Helping myself to achieve my goals	5.87	4.81	−1.06
Saving money or cost-cutting (doing my job with less money)	5.74	4.73	−1.01
Saving time (doing my job in less time)	5.68	4.55	−1.13
Putting the interests of others before my own	5.51	4.54	−0.97
Relying only on myself for help (being independent)	5.30	4.68	−0.62
Spending more time at my job	5.23	4.38	−0.85
Competing with my peers	5.17	4.79	−0.38
Being intellectual	5.08	4.36	−0.72
Socializing with peers after work	4.70	4.27	−0.43
Relying on my peers when I need help (being dependent)	4.69	4.19	−0.50
Marrying	4.21	4.21	0.00
Having children	4.08	4.10	+0.02
Acting like an individual first and a member of my unit second	2.64	3.35	+0.71

NOTE: "How Desirable" is rated on a scale from 1 ("extremely undesirable") to 7 ("extremely desirable"). "How Rewarded" is rated on a scale from 1 ("punished severely") to 7 ("rewarded a lot"). On both scales, 4 is a neutral midpoint.

Table A.3

Mean Size of Effect on Work of Categories of Rewards and Punishments

Categories of Rewards and Punishments	Size of Effect on Work
Whether you felt like a valued and respected member of your unit	3.24
How much you respected your direct supervisors and commanding officers	3.08
Your job assignment	3.07
Amount of praise you received for desirable behavior	3.03
How fairly you were treated in promotion decisions	2.95
Amount of criticism you received for undesirable behavior	2.89
Your working conditions	2.89
How much you liked working and socializing with your peers	2.81
How much choice you were given about your job assignment	2.63
How much choice you were given about where you were stationed and for how long	2.58
Your base location	2.52
Your pay	2.27
Your benefits	2.22

NOTE: All mean differences greater than .13 are significant at p<.05. Scale: 1 = "had no effect on your work"; 2 = "had a small effect on your work"; 3 = "had a moderate effect on your work"; 4 = "had a large effect on your work."

Table A.4

Mean Size and Direction of Effect on Work of Levels of Rewards and Punishments

Levels of Rewards and Punishments	Size and Direction of Effect on Work
Receiving a promotion you feel you deserve	5.93
Compared to being assigned to an average job, being assigned to your favorite realistic job	5.76
Being given more choice about where you are stationed and for how long	5.66
Being given more choice about your job assignment	5.62
Receiving more praise for desirable behavior from your superiors	5.39
Receiving $200/month more in basic pay	5.37
Compared to being stationed at an average location, being stationed at a very desirable location	5.28
Receiving $100/month more in basic pay	4.85
Receiving less criticism for undesirable behavior from your superiors	4.10
Receiving more criticism for undesirable behavior from your superiors	3.75
Receiving less praise for desirable behavior from your superiors	3.54
Compared to being stationed at an average location, being stationed at a very undesirable location	2.93
Receiving $100/month less in basic pay	2.79
Compared to being assigned to an average job, being assigned to your worst realistic job	2.74
Not receiving a promotion you feel you deserve	2.59
Receiving $200/month less in basic pay	2.38

NOTE: All mean differences greater than .17 are significant at p<.05. Scale: 1 = "much worse"; 2 = "moderately worse"; 3 = "slightly worse"; 4 = "no effect"; 5 = "slightly better"; 6 = "moderately better"; 7 = "much better."

74 Differentiation in Military Human Resource Management

Fairness and Unfairness of Military Compensation

In this section of the survey, we wanted to learn how fair military personnel perceive the level and distribution of military pay and benefits to be. We first asked military personnel three general questions, one about overall satisfaction with the amount of their pay and benefits and two about perceived fairness of their pay and benefits compared with that of their peers in and out of the military.

For the first question, we asked them to indicate "Which statement best characterizes your thinking about your level of pay and benefits?" Their four options were (1) "Overall, I am very unsatisfied. I think I receive way too little pay and way too few benefits." (2) "Overall, I am unsatisfied. I think I receive too little pay and too few benefits." (3) "Overall, I am satisfied. I think I receive about the right amount of pay and benefits." (4) "Overall, I am very satisfied. I think I receive a generous amount of pay and benefits." The results are reported in Figure A.1.

For the second and third questions, we asked them to complete the following statement: "Compared to my peers *in the military [outside the military]*, I think my level of pay and benefits is..." Their five options were: "way too little," "too little," "about right," too much," and "way too much." The results are shown in Figure A.2.

In addition, as reported in Table A.5, we provided military personnel with a list of 16 factors and asked them to indicate "which factors *do affect* and *should affect* total compensation (includes all forms of compensation such as basic pay, special pay, bonuses, housing, food, health care, and retirement)." They answered "yes" or "no" to each question. We also created a difference score—simply the difference between the proportions of respondents who answered "yes" to the *should affect* and

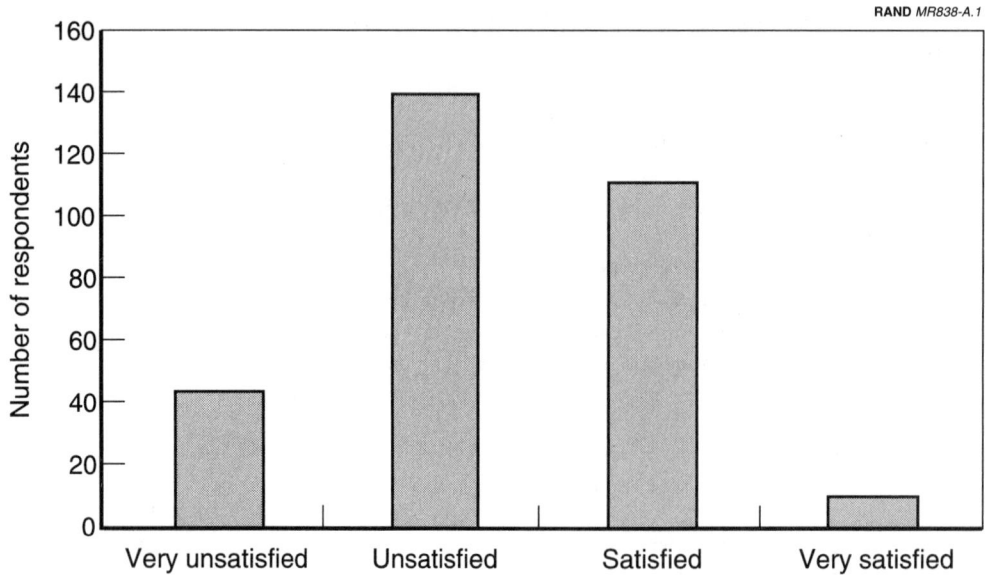

Figure A.1—Degree of Satisfaction with the Level of Military Pay and Benefits

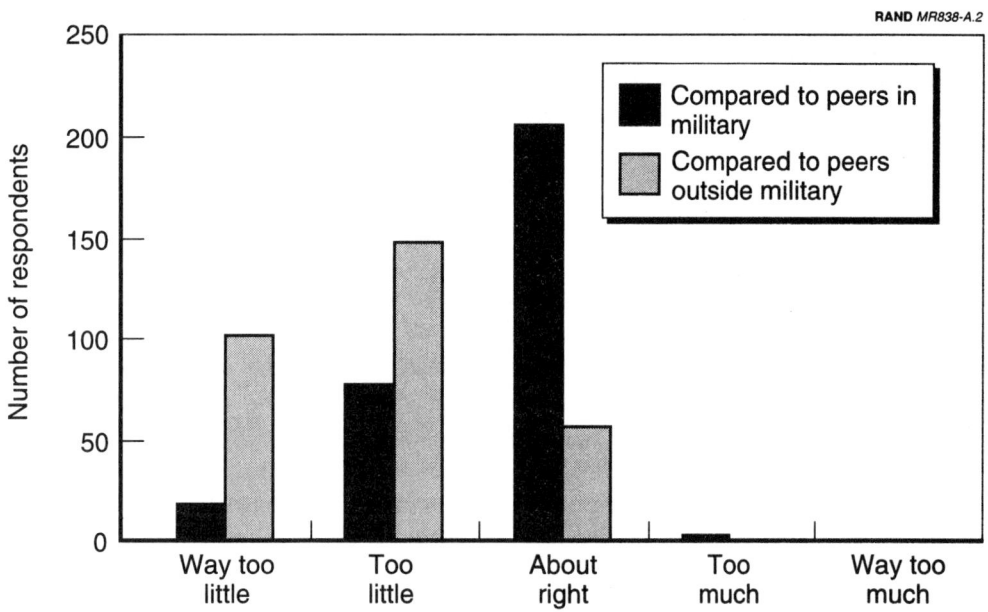

Table A.5

Proportion of Military Personnel Who Believe That Potential Compensation Factors Should and Do Affect Military Compensation

Factor	Should Affect?	Does Affect?	Difference
Rank/rate	98%	100%	0.02
Cost of living in your area	98%	87%	−0.11
Years of service	93%	99%	0.06
Being at war vs. at peace	84%	75%	−0.09
Officer vs. Enlisted	72%	90%	0.18
Individual performance (how well you do your job)	69%	31%	−0.38
Effort (how hard you work)	62%	26%	−0.36
Occupation	57%	55%	−0.02
Marital status	55%	85%	0.30
Education	53%	68%	−0.15
Knowledge (as measured by tests)	52%	30%	−0.22
Number of dependents	48%	56%	0.08
Service	42%	41%	−0.01
Physical fitness	41%	25%	−0.16
Group performance (how well your unit does its job)	36%	10%	−0.26
Accession source	19%	27%	0.08

does affect questions—which provides some clue about where discrepancies are perceived between what should and what does affect compensation. A large positive difference score may indicate a factor that should not affect but that does affect compensation. A large negative difference score may indicate a factor that should affect but that does not affect compensation. We put in boldface type any difference score that was above an absolute value of .20.

Attitudes Toward Innovations in Human Resource Management

In the final section of the survey, we wanted to learn the attitudes of military personnel toward innovative ways of compensating, managing, and organizing people that are currently being used in many private-sector organizations. We asked them to indicate their attitudes toward eight HRM techniques on a 7-point scale (with 1 = "extremely unappealing," 4 = "indifferent," and 7 = "extremely appealing").

The results are displayed in Table A.6.

Differentiation of Survey Results by Community

The survey findings reported above do not take into account differences that may exist across communities of the military. Here, we examine differences across the four services, combat and noncombat communities, officer and enlisted communities, and senior and junior personnel communities.

We conducted an exploratory analysis of community differences on the over 100 dependent measures taken in the survey. Because of the large number of statistical tests and the exploratory nature of this exercise, we adopted strict standards for concluding that a community difference exists. This strategy helps to reduce family-wise

Table A.6

Mean Attitudes of Military Personnel Toward Human Resource Management Techniques

Human Resource Management Technique	Attitude
Individual performance as a basis for promotion	5.96
Pay for human capital development (training, education, experience, physical fitness, etc.)	4.67
Pay for individual performance	4.55
Team or organizational performance as a basis for promotion	4.07
Pay for team or organizational performance (gainsharing or goalsharing)	4.04
Self-managed teams (teams that manage their own task and interpersonal processes, without first-line supervisors)	3.93
360-degree evaluation system (evaluation by peers and subordinates as well as by supervisors)	3.55
Lateral entries (direct appointment of qualified individuals at middle or senior grades)	2.68

NOTE: The HRM techniques were rated on a scale from 1 ("extremely unappealing") to 7 ("extremely appealing"), with 4 as the midpoint ("indifferent").

Type I error, which is simply the rate of Type I error (falsely concluding that a difference exists) over the analysis as a whole (Keppel and Zedeck, 1989).

To look for differences among the services, we conducted over 100 ANOVAs, one for each dependent measure, using service as a categorical dummy variable. Only when the omnibus F-test was significant at the p<.01 level (slightly stricter than the standard p<.05 level) did we proceed to look for the source of service differences. All post hoc single-degree-of-freedom comparisons were evaluated using the Tukey test, which reduces family-wise Type I error.

To look for differences between the combat and noncombat communities, we calculated over 100 Pearson correlation coefficients, one for each dependent measure's relation to the combat variable. The significance of each correlation was assessed using the strict Bonferroni test for significant differences, which reduces family-wise Type I error. We used the same procedure to test for differences between the officer/enlisted communities and junior/senior communities.

The results are shown in Tables A.7 through A.13.

Table A.7

Mean Desirability of Behaviors, by Community

Behavior	By Service	Combat vs. Noncombat	Officer vs. Enlisted	Junior vs. Senior
Obeying orders of my superiors	A = 6.54 N = 6.70 AF = 6.78 M = 6.88	C = 6.72 N = 6.75	O = 6.86 E = 6.61	J = 6.76 S = 6.70
Helping my unit to achieve its goals	A = 6.67 N = 6.70 AF = 6.62 M = 6.81	C = 6.75 N = 6.68	O = 6.83 E = 6.60	J = 6.72 S = 6.70
Respecting authority	A = 6.47 N = 6.60 AF = 6.72 M = 6.79	C = 6.61 N = 6.69	O = 6.78 E = 6.52	J = 6.70 S = 6.60
Taking initiative; acting like a leader	A = 6.43 N = 6.55 AF = 6.41 M = 6.67	C = 6.49 N = 6.57	O = 6.57 E = 6.50	J = 6.51 S = 6.56
Having a good attitude at work	A = 6.41 N = 6.19 AF = 6.40 M = 6.46	C = 6.30 N = 6.44	O = 6.44 E = 6.31	J = 6.41 S = 6.32
Working cooperatively with my peers	A = 6.01 N = 6.20 AF = 6.24 M = 6.45	C = 6.19 N = 6.29	O = 6.25 E = 6.25	J = 6.22 S = 6.28
Conforming to the norms and values of the military	A = 6.00 N = 6.20 AF = 6.28 M = 6.47[c]	C = 6.16 N = 6.34	O = 6.35 E = 6.16	J = 6.22 S = 6.30

Table A.7—continued

Behavior	By Service	Combat vs. Noncombat	Officer vs. Enlisted	Junior vs. Senior
Being more productive at my job	A = 6.04 N = 6.20 AF = 6.18 M = 6.43	C = 6.20 N = 6.26	O = 6.37 E = 6.10	J = 6.28 S = 6.18
Exerting more effort at my job	A = 6.20 N = 6.15 AF = 5.93 M = 6.20	C = 6.14 N = 6.14	O = 6.26 E = 6.01	J = 6.26 S = 6.00
Making fewer mistakes at my job	A = 5.86 N = 6.15 AF = 6.30 M = 6.15	C = 6.18 N = 6.05	O = 6.31 E = 5.91	J = 6.15 S = 6.06
Maintaining physical fitness	A = 6.39 N = 5.85 AF = 5.60 M = 6.39[a,b,e,f]	C = 6.04 N = 6.16	O = 5.97 E = 6.23	J = 6.13 S = 6.08
Thinking creatively (finding new solutions to old problems)	A = 5.73 N = 5.97 AF = 5.88 M = 6.43[c,f]	C = 5.98 N = 6.10	O = 6.11 E = 5.97	J = 5.97 S = 6.11
Helping my peers to achieve their goals	A = 5.81 N = 5.95 AF = 5.91 M = 6.30	C = 6.01 N = 6.03	O = 5.95 E = 6.10	J = 6.07 S = 5.97
Helping myself to achieve my goals	A = 5.90 N = 5.77 AF = 5.74 M = 5.98	C = 5.79 N = 5.92	O = 5.71 E = 6.01	J = 5.98 S = 5.73
Saving money or cost-cutting (doing my job with less money)	A = 5.58 N = 5.78 AF = 5.62 M = 5.82	C = 5.76 N = 5.68	O = 5.78 E = 5.65	J = 5.75 S = 5.68
Saving time (doing my job in less time)	A = 5.79 N = 5.47 AF = 5.75 M = 5.70	C = 5.65 N = 5.69	O = 5.64 E = 5.70	J = 5.75 S = 5.59
Putting the interests of others before my own	A = 5.70 N = 5.24 AF = 5.33 M = 5.70	C = 5.61 N = 5.43	O = 5.70 E = 5.32	J = 5.35 S = 5.69
Relying only on myself for help (being independent)	A = 4.90 N = 5.47 AF = 5.24 M = 5.58	C = 5.30 N = 5.35	O = 5.31 E = 5.35	J = 5.22 S = 5.44
Spending more time at my job	A = 5.21 N = 5.18 AF = 5.26 M = 5.33	C = 5.39 N = 5.13	O = 5.28 E = 5.22	J = 5.33 S = 5.17
Competing with my peers	A = 5.16 N = 4.82 AF = 5.12 M = 5.51	C = 5.28 N = 5.08	O = 5.12 E = 5.23	J = 5.10 S = 5.26

Table A.7—continued

Behavior	By Service	Combat vs. Noncombat	Officer vs. Enlisted	Junior vs. Senior
Being intellectual	A = 5.02 N = 5.15 AF = 5.06 M = 5.08	C = 5.08 N = 5.07	O = 5.09 E = 5.06	J = 5.12 S = 5.03
Socializing with peers after work	A = 4.54 N = 4.53 AF = 5.20 M = 4.65[b,d,f]	C = 4.91 N = 4.49	O = 5.00 E = 4.39[g]	J = 4.83 S = 4.55
Relying on my peers when I need help (being dependent)	A = 4.68 N = 4.61 AF = 4.95 M = 4.61	C = 4.73 N = 4.66	O = 4.61 E = 4.78	J = 4.54 S = 4.86
Marrying	A = 4.24 N = 4.16 AF = 4.43 M = 4.09	C = 4.19 N = 4.22	O = 4.23 E = 4.18	J = 4.22 S = 4.19
Having children	A = 4.16 N = 4.07 AF = 4.20 M = 3.97	C = 4.13 N = 4.04	O = 4.12 E = 4.05	J = 4.07 S = 4.10
Acting like an individual first and a member of my unit second	A = 2.71 N = 2.92 AF = 2.90 M = 2.22	C = 2.64 N = 2.65	O = 2.31 E = 2.97	J = 2.52 S = 2.78

NOTES: [a]Army and Navy differ, [b]Army and Air Force differ, [c]Army and Marines differ, [d]Navy and Air Force differ, [e]Navy and Marines differ, [f]Air Force and Marines differ, [g]Combat and Noncombat, Officer and Enlisted, or Junior and Senior differ.

The behavior was rated on a scale from 1 ("extremely undesirable") to 7 ("extremely desirable"), with 4 as the midpoint ("neutral").

Table A.8

Mean Rewardedness of Behaviors, by Community

Behavior	By Service	Combat vs. Noncombat	Officer vs. Enlisted	Junior vs. Senior
Obeying orders of my superiors	A = 4.67 N = 4.60 AF = 4.47 M = 4.86	C = 4.64 N = 4.71	O = 4.73 E = 4.62	J = 4.61 S = 4.74
Helping my unit to achieve its goals	A = 5.51 N = 5.30 AF = 5.18 M = 5.59	C = 5.46 N = 5.38	O = 5.58 E = 5.26	J = 5.43 S = 5.41
Respecting authority	A = 4.57 N = 4.32 AF = 4.47 M = 4.86	C = 4.50 N = 4.65	O = 4.66 E = 4.49	J = 4.47 S = 4.70
Taking initiative; acting like a leader	A = 5.41 N = 5.38 AF = 5.20 M = 5.77[f]	C = 5.42 N = 5.53	O = 5.64 E = 5.32	J = 5.48 S = 5.48

Table A.8—continued

Behavior	By Service	Combat vs. Noncombat	Officer vs. Enlisted	Junior vs. Senior
Having a good attitude at work	A = 4.76 N = 4.82 AF = 4.66 M = 4.88	C = 4.76 N = 4.83	O = 4.81 E = 4.78	J = 4.77 S = 4.82
Working cooperatively with my peers	A = 4.75 N = 4.55 AF = 4.58 M = 4.73	C = 4.62 N = 4.70	O = 4.70 E = 4.62	J = 4.64 S = 4.68
Conforming to the norms and values of the military	A = 4.58 N = 4.43 AF = 4.60 M = 5.20	C = 4.51 N = 4.72	O = 4.70 E = 4.54	J = 4.57 S = 4.67
Being more productive at my job	A = 4.94 N = 4.97 AF = 4.86 M = 5.22	C = 4.97 N = 5.07	O = 5.16 E = 4.88	J = 5.02 S = 5.02
Exerting more effort at my job	A = 4.77 N = 4.86 AF = 4.71 M = 5.02	C = 4.84 N = 4.89	O = 4.91 E = 4.81	J = 4.91 S = 4.82
Making fewer mistakes at my job	A = 4.41 N = 4.84 AF = 4.61 M = 4.80	C = 4.76 N = 4.61	O = 4.87 E = 4.49	J = 4.75 S = 4.60
Maintaining physical fitness	A = 5.16 N = 4.59 AF = 4.13 M = 5.03[a,b,d,e,f]	C = 4.70 N = 4.85	O = 4.74 E = 4.81	J = 4.82 S = 4.73
Thinking creatively (finding new solutions to old problems)	A = 5.13 N = 5.18 AF = 4.80 M = 5.54[f]	C = 5.22 N = 5.21	O = 5.33 E = 5.09	J = 5.20 S = 5.22
Helping my peers to achieve their goals	A = 4.80 N = 4.77 AF = 4.63 M = 5.08	C = 4.92 N = 4.78	O = 4.84 E = 4.86	J = 4.89 S = 4.80
Helping myself to achieve my goals	A = 4.80 N = 4.86 AF = 4.61 M = 4.86	C = 4.84 N = 4.76	O = 4.66 E = 4.94	J = 4.80 S = 4.80
Saving money or cost-cutting (doing my job with less money)	A = 4.51 N = 4.78 AF = 4.57 M = 4.88	C = 4.70 N = 4.72	O = 4.71 E = 4.71	J = 4.79 S = 4.62
Saving time (doing my job in less time)	A = 4.60 N = 4.52 AF = 4.50 M = 4.54	C = 4.53 N = 4.55	O = 4.56 E = 4.52	J = 4.60 S = 4.48

Table A.8—continued

Behavior	By Service	Combat vs. Noncombat	Officer vs. Enlisted	Junior vs. Senior
Putting the interests of others before my own	A = 4.47 N = 4.36 AF = 4.46 M = 4.75	C = 4.60 N = 4.47	O = 4.59 E = 4.47	J = 4.54 S = 4.51
Relying only on myself for help (being independent)	A = 4.61 N = 4.81 AF = 4.33 M = 4.80	C = 4.72 N = 4.62	O = 4.69 E = 4.65	J = 4.66 S = 4.68
Spending more time at my job	A = 4.20 N = 4.38 AF = 4.36 M = 4.46	C = 4.36 N = 4.36	O = 4.48 E = 4.23	J = 4.36 S = 4.36
Competing with my peers	A = 4.76 N = 4.82 AF = 4.61 M = 4.89	C = 4.87 N = 4.71	O = 4.88 E = 4.70	J = 4.76 S = 4.82
Being intellectual	A = 4.24 N = 4.45 AF = 4.44 M = 4.35	C = 4.35 N = 4.38	O = 4.28 E = 4.45	J = 4.40 S = 4.33
Socializing with peers after work	A = 4.14 N = 4.33 AF = 4.50 M = 4.17	C = 4.39 N = 4.16	O = 4.45 E = 4.08	J = 4.33 S = 4.20
Relying on my peers when I need help (being dependent)	A = 4.33 N = 4.06 AF = 4.18 M = 4.21	C = 4.22 N = 4.17	O = 4.15 E = 4.24	J = 4.14 S = 4.25
Marrying	A = 4.11 N = 4.29 AF = 4.20 M = 4.24	C = 4.30 N = 4.13	O = 4.19 E = 4.24	J = 4.22 S = 4.21
Having children	A = 4.07 N = 4.15 AF = 4.04 M = 4.13	C = 4.20 N = 4.01	O = 4.06 E = 4.15	J = 4.15 S = 4.06
Acting like an individual first and a member of my unit second	A = 3.30 N = 3.53 AF = 3.50 M = 3.17	C = 3.38 N = 3.33	O = 3.04 E = 3.67[g]	J = 3.27 S = 3.44

NOTES: [a]Army and Navy differ, [b]Army and Air Force differ, [c]Army and Marines differ, [d]Navy and Air Force differ, [e]Navy and Marines differ, [f]Air Force and Marines differ, [g]Combat and Noncombat, Officer and Enlisted, or Junior and Senior differ.

The rewards for each behavior were rated on a scale from 1 ("punished severely") to 7 ("rewarded a lot"), with 4 as the midpoint ("neither punished nor rewarded").

Table A.9

Mean Size of Effect on Work of Categories of Rewards and Punishments, by Community

Categories of Rewards and Punishments	By Service	Combat vs. Noncombat	Officer vs. Enlisted	Junior vs. Senior
Whether you felt like a valued and respected member of your unit	A = 3.24 N = 3.43 AF = 3.17 M = 3.10	C = 3.33 N = 3.14	O = 3.37 E = 3.09	J = 3.19 S = 3.27
How much you respected your direct supervisors and commanding officers	A = 2.97 N = 3.29 AF = 3.12 M = 3.00	C = 3.06 N = 3.12	O = 3.25 E = 2.93	J = 3.11 S = 3.06
Your job assignment	A = 3.20 N = 3.35 AF = 2.79 M = 2.87[d,e]	C = 3.14 N = 2.98	O = 3.07 E = 3.04	J = 3.05 S = 3.06
Amount of praise you received for desirable behavior	A = 3.10 N = 3.12 AF = 2.88 M = 2.92	C = 3.04 N = 2.98	O = 3.10 E = 2.92	J = 3.08 S = 2.93
How fairly you were treated in promotion decisions	A = 3.13 N = 3.03 AF = 2.88 M = 2.81	C = 3.09 N = 2.83	O = 2.82 E = 3.09	J = 2.88 S = 3.04
Amount of criticism you received for undesirable behavior	A = 2.81 N = 2.96 AF = 2.83 M = 2.90	C = 2.96 N = 2.81	O = 2.97 E = 2.79	J = 2.89 S = 2.87
Your working conditions	A = 3.17 N = 3.10 AF = 2.69 M = 2.70[b,c,e]	C = 2.92 N = 2.90	O = 2.71 E = 3.11	J = 3.04 S = 2.77
How much you liked working and socializing with your peers	A = 2.91 N = 2.93 AF = 2.67 M = 2.73	C = 2.96 N = 2.68	O = 2.86 E = 2.76	J = 2.87 S = 2.75
How much choice you were given about your job assignment	A = 2.83 N = 2.60 AF = 2.67 M = 2.42	C = 2.63 N = 2.59	O = 2.67 E = 2.55	J = 2.63 S = 2.59
How much choice you were given about where you were stationed and for how long	A = 2.70 N = 2.65 AF = 2.65 M = 2.39	C = 2.59 N = 2.57	O = 2.52 E = 2.63	J = 2.61 S = 2.55
Your base location	A = 2.71 N = 2.70 AF = 2.40 M = 2.33	C = 2.61 N = 2.42	O = 2.39 E = 2.63	J = 2.51 S = 2.51

Table A.9—continued

Levels of Rewards and Punishments	By Service	Combat vs. Noncombat	Officer vs. Enlisted	Junior vs. Senior
Your pay	A = 2.50 N = 2.46 AF = 2.12 M = 1.98[c,e]	C = 2.28 N = 2.23	O = 2.08 E = 2.43	J = 2.18 S = 2.32
Your benefits	A = 2.43 N = 2.42 AF = 2.11 M = 1.94[c,e]	C = 2.27 N = 2.15	O = 2.07 E = 2.35	J = 2.13 S = 2.29

NOTES: [a]Army and Navy differ, [b]Army and Air Force differ, [c]Army and Marines differ, [d]Navy and Air Force differ, [e]Navy and Marines differ, [f]Air Force and Marines differ, [g]Combat and Noncombat, Officer and Enlisted, or Junior and Senior differ.

The influence of the factors was rated on a scale from 1 ("had no effect on your work") to 4 ("had a large effect on your work").

Table A.10

Mean Size and Direction of Effect on Work of Levels of Rewards and Punishments, by Community

Levels of Rewards and Punishments	By Service	Combat vs. Noncombat	Officer vs. Enlisted	Junior vs. Senior
Receiving a promotion you feel you deserve	A = 6.26 N = 5.96 AF = 5.86 M = 5.72	C = 5.99 N = 5.89	O = 5.71 E = 6.16	J = 5.97 S = 5.89
Compared to being assigned to an average job, being assigned to your favorite realistic job	A = 5.90 N = 5.87 AF = 5.62 M = 5.62	C = 5.82 N = 5.68	O = 5.66 E = 5.84	J = 5.85 S = 5.65
Being given more choice about where you are stationed and for how long	A = 6.00 N = 5.78 AF = 5.58 M = 5.28[c]	C = 5.61 N = 5.66	O = 5.50 E = 5.78	J = 5.67 S = 5.60
Being given more choice about your job assignment	A = 5.96 N = 5.62 AF = 5.61 M = 5.31[c]	C = 5.71 N = 5.50	O = 5.46 E = 5.75	J = 5.60 S = 5.60
Receiving more praise for desirable behavior from your superiors	A = 5.51 N = 5.39 AF = 5.38 M = 5.28	C = 5.47 N = 5.30	O = 5.42 E = 5.34	J = 5.46 S = 5.30
Receiving $200/month more in basic pay	A = 5.79 N = 5.26 AF = 5.34 M = 5.12[a,c]	C = 5.26 N = 5.44	O = 4.91 E = 5.81[g]	J = 5.47 S = 5.23
Compared to being stationed at an average location, being stationed at a very desirable location	A = 5.49 N = 5.45 AF = 5.14 M = 5.05	C = 5.30 N = 5.25	O = 5.08 E = 5.46	J = 5.49 S = 5.04

Table A.10—continued

Levels of Rewards and Punishments	By Service	Combat vs. Noncombat	Officer vs. Enlisted	Junior vs. Senior
Receiving $100/month more in basic pay	A = 5.10 N = 4.81 AF = 4.77 M = 4.72	C = 4.83 N = 4.86	O = 4.54 E = 5.14[g]	J = 4.98 S = 4.70
Receiving less criticism for undesirable behavior from your superiors	A = 4.21 N = 4.14 AF = 3.93 M = 4.12	C = 4.16 N = 4.07	O = 4.05 E = 4.16	J = 4.07 S = 4.16
Receiving more criticism for undesirable behavior from your superiors	A = 3.71 N = 3.72 AF = 3.82 M = 3.84	C = 3.89 N = 3.67	O = 4.04 E = 3.51	J = 3.85 S = 3.69
Receiving less praise for desirable behavior from your superiors	A = 3.41 N = 3.38 AF = 3.46 M = 3.83[c,e]	C = 3.59 N = 3.50	O = 3.54 E = 3.56	J = 3.47 S = 3.63
Compared to being stationed at an average location, being stationed at a very undesirable location	A = 2.64 N = 2.70 AF = 3.21 M = 3.23[b,c,e]	C = 2.89 N = 3.01	O = 3.16 E = 2.75	J = 2.87 S = 3.04
Receiving $100/month less in basic pay	A = 2.31 N = 2.65 AF = 3.05 M = 3.08[b,c]	C = 2.79 N = 2.77	O = 3.14 E = 2.43[g]	J = 2.69 S = 2.88
Compared to being assigned to an average job, being assigned to your worst realistic job	A = 2.42 N = 2.61 AF = 2.93 M = 3.00	C = 2.71 N = 2.79	O = 2.90 E = 2.60	J = 2.63 S = 2.88
Not receiving a promotion you feel you deserve	A = 2.34 N = 2.64 AF = 2.41 M = 2.88	C = 2.59 N = 2.61	O = 2.53 E = 2.67	J = 2.48 S = 2.73
Receiving $200/month less in basic pay	A = 1.86 N = 2.27 AF = 2.61 M = 2.71[b,c]	C = 2.46 N = 2.29	O = 2.81 E = 1.94[g]	J = 2.27 S = 2.49

NOTES: [a]Army and Navy differ, [b]Army and Air Force differ, [c]Army and Marines differ, [d]Navy and Air Force differ, [e]Navy and Marines differ, [f]Air Force and Marines differ, [g]Combat and Noncombat, Officer and Enlisted, or Junior and Senior differ.

The effect each reward or punishment would have on the quality and quantity of work was rated on a scale from 1 ("much worse") to 7 ("much better"), with 4 as the midpoint ("no effect").

Table A.11

Proportion of Military Personnel Who Believe That Potential Compensation Factors Should Affect Military Compensation, by Community

Factor	By Service	Combat vs. Noncombat	Officer vs. Enlisted	Junior vs. Senior
Rank/rate	A = 98% N = 100% AF = 96% M = 98%	C = 99% N = 97%	O = 100% E = 96%	J = 98% S = 98%
Cost of living in your area	A = 100% N = 98% AF = 95% M = 97%	C = 98% N = 96%	O = 93% E = 98%	J = 97% S = 98%
Years of service	A = 97% N = 94% AF = 84% M = 95%	C = 96% N = 90%	O = 95% E = 90%	J = 90% S = 96%
Being at war vs. at peace	A = 97% N = 81% AF = 80% M = 82%	C = 87% N = 82%	O = 83% E = 86%	J = 82% S = 87%
Officer vs. Enlisted	A = 68% N = 82% AF = 73% M = 67%	C = 75% N = 69%	O = 90% E = 53%[g]	J = 71% S = 72%
Individual performance (how well you do your job)	A = 56% N = 76% AF = 75% M = 68%	C = 66% N = 71%	O = 71% E = 66%	J = 72% S = 65%
Effort (how hard you work)	A = 55% N = 75% AF = 73% M = 53%[e]	C = 59% N = 66%	O = 56% E = 69%	J = 54% S = 62%
Occupation	A = 70% N = 72% AF = 53% M = 36%[c,e]	C = 64% N = 49%	O = 60% E = 52%	J = 62% S = 50%
Marital status	A = 83% N = 43% AF = 49% M = 50%[a,b,c]	C = 59% N = 52%	O = 52% E = 59%	J = 57% S = 54%
Education	A = 52% N = 48% AF = 62% M = 46%	C = 46% N = 61%	O = 56% E = 51%	J = 60% S = 47%
Knowledge (as measured by tests)	A = 52% N = 51% AF = 61% M = 50%	C = 47% N = 59%	O = 54% E = 52%	J = 61% S = 45%
Number of dependents	A = 74% N = 40% AF = 43% M = 42%[a,b,c]	C = 49% N = 49%	O = 44% E = 54%	J = 57% S = 40%

Table A.11—continued

Factor	By Service	Combat vs. Noncombat	Officer vs. Enlisted	Junior vs. Senior
Service	A = 64% N = 48% AF = 33% M = 30%[b,c]	C = 55% N = 32%[g]	O = 35% E = 52%	J = 47% S = 39%
Physical fitness	A = 44% N = 44% AF = 40% M = 39%	C = 39% N = 44%	O = 39% E = 44%	J = 43% S = 40%
Group performance (how well your unit does its job)	A = 25% N = 48% AF = 37% M = 30%	C = 34% N = 35%	O = 44% E = 25%	J = 42% S = 28%
Accession source	A = 22% N = 23% AF = 6% M = 21%	C = 20% N = 18%	O = 10% E = 30%[g]	J = 22% S = 16%

NOTES: [a]Army and Navy differ, [b]Army and Air Force differ, [c]Army and Marines differ, [d]Navy and Air Force differ, [e]Navy and Marines differ, [f]Air Force and Marines differ, [g]Combat and Noncombat, Officer and Enlisted, or Junior and Senior differ.

Table A.12

Proportion of Military Personnel Who Believe That Potential Compensation Factors Do Affect Military Compensation, by Community

Factor	By Service	Combat vs. Noncombat	Officer vs. Enlisted	Junior vs. Senior
Rank/rate	A = 100% N = 100% AF = 100% M = 99%	C = 99% N = 100%	O = 99% E = 100%	J = 100% S = 99%
Cost of living in your area	A = 85% N = 93% AF = 80% M = 89%	C = 88% N = 87%	O = 91% E = 83%	J = 88% S = 87%
Years of service	A = 100% N = 100% AF = 96% M = 99%	C = 98% N = 99%	O = 99% E = 98%	J = 100% S = 97%
Being at war vs. at peace	A = 85% N = 79% AF = 58% M = 79%[b,d,f]	C = 81% N = 72%	O = 76% E = 76%	J = 76% S = 76%
Officer vs. Enlisted	A = 87% N = 91% AF = 94% M = 90%	C = 94% N = 87%	O = 96% E = 85%[g]	J = 92% S = 89%
Individual performance (how well you do your job)	A = 33% N = 29% AF = 25% M = 36%	C = 30% N = 32%	O = 24% E = 39%	J = 32% S = 31%

Table A.12—continued

Factor	By Service	Combat vs. Noncombat	Officer vs. Enlisted	Junior vs. Senior
Effort (how hard you work)	A = 34% N = 29% AF = 13% M = 26%	C = 26% N = 26%	O = 18% E = 34%	J = 23% S = 29%
Occupation	A = 57% N = 65% AF = 55% M = 43%	C = 62% N = 47%	O = 70% E = 37%[g]	J = 52% S = 56%
Marital status	A = 88% N = 85% AF = 78% M = 89%	C = 88% N = 83%	O = 89% E = 82%	J = 85% S = 86%
Education	A = 46% N = 40% AF = 33% M = 39%	C = 40% N = 39%	O = 36% E = 44%	J = 40% S = 39%
Knowledge (as measured by tests)	A = 16% N = 46% AF = 36% M = 22%	C = 35% N = 25%	O = 18% E = 42%[g]	J = 36% S = 24%
Number of dependents	A = 66% N = 50% AF = 54% M = 52%	C = 56% N = 54%	O = 59% E = 51%	J = 61% S = 49%
Service	A = 55% N = 45% AF = 37% M = 31%	C = 48% N = 35%	O = 35% E = 47%	J = 45% S = 37%
Physical fitness	A = 35% N = 27% AF = 7% M = 30%[b,f]	C = 25% N = 26%	O = 15% E = 37%[g]	J = 28% S = 22%
Group performance (how well your unit does its job)	A = 10% N = 12% AF = 7% M = 10%	C = 11% N = 9%	O = 12% E = 8%	J = 12% S = 8%
Accession source	A = 24% N = 38% AF = 15% M = 30%	C = 31% N = 25%	O = 21% E = 35%	J = 31% S = 25%

NOTES: [a]Army and Navy differ, [b]Army and Air Force differ, [c]Army and Marines differ, [d]Navy and Air Force differ, [e]Navy and Marines differ, [f]Air Force and Marines differ, [g]Combat and Noncombat, Officer and Enlisted, or Junior and Senior differ.

Table A.13

Mean Attitudes of Military Personnel Toward Human Resource Management Techniques, by Community

Human Resource Management Technique	By Service	Combat vs. Noncombat	Officer vs. Enlisted	Junior vs. Senior
Individual performance as a basis for promotion	A = 5.77 N = 6.03 AF = 5.86 M = 6.16	C = 6.02 N = 5.93	O = 6.12 E = 5.83	J = 6.01 S = 5.93
Pay for human capital development (training, education, experience, physical fitness, etc.)	A = 4.34 N = 4.84 AF = 5.27 M = 4.41[b,f]	C = 4.50 N = 4.84	O = 4.70 E = 4.64	J = 4.88 S = 4.46
Pay for individual performance	A = 4.20 N = 4.94 AF = 5.11 M = 4.02[b,e,f]	C = 4.22 N = 4.79	O = 4.53 E = 4.49	J = 4.83 S = 4.17
Team or organizational performance as a basis for promotion	A = 3.74 N = 4.66 AF = 4.00 M = 3.84	C = 4.11 N = 3.99	O = 4.32 E = 3.78	J = 4.16 S = 3.93
Pay for team or organizational performance (gainsharing or goalsharing)	A = 3.69 N = 4.70 AF = 4.09 M = 3.59[a,e]	C = 4.05 N = 3.93	O = 4.10 E = 3.87	J = 4.20 S = 3.76
Self-managed teams (teams that manage their own task and interpersonal processes, without first-line supervisors)	A = 3.14 N = 4.09 AF = 4.68 M = 3.75[a,b,f]	C = 4.03 N = 3.74	O = 3.65 E = 4.11	J = 4.12 S = 3.63
360-degree evaluation system (evaluation by peers and subordinates, as well as by supervisors)	A = 3.49 N = 3.33 AF = 4.07 M = 3.33	C = 3.64 N = 3.40	O = 3.41 E = 3.62	J = 4.18 S = 2.84[g]
Lateral entries (direct appointment of qualified individuals at middle or senior grades)	A = 2.85 N = 2.73 AF = 3.36 M = 2.08[c,f]	C = 2.60 N = 2.75	O = 2.70 E = 2.66	J = 3.11 S = 2.24

NOTES: [a]Army and Navy differ, [b]Army and Air Force differ, [c]Army and Marines differ, [d]Navy and Air Force differ, [e]Navy and Marines differ, [f]Air Force and Marines differ, [g]Combat and Noncombat, Officer and Enlisted, or Junior and Senior differ.

Perceptions of different ways of compensating, managing, and organizing people were rated on a scale from 1 ("extremely unappealing") to 7 ("extremely appealing"), with 4 as the midpoint ("indifferent").

Exhibit A.1

INTRODUCTORY REMARKS FOR FOCUS GROUPS

Hello, I'm Dr. Mark Spranca, a researcher at RAND. I will be leading today's discussion. I would like to thank you for coming. We're looking forward to hearing your thoughts and feelings about the military's compensation system. Before we get started, I would like to introduce the rest of us.

> This is <u>name</u>, a researcher at RAND. (S)He will be assisting me in today's discussion.
> This is <u>name and affiliation</u>. (S)He is here to take notes.
> This is <u>name</u>, a member of the QRMC staff. (S)He is here to observe today's discussion.
> We also have <u>name and affiliation of anyone else present,</u> who will be observing today's discussion.

Let me tell you a little about RAND and why we have asked you to participate.

RAND is a nonprofit research organization. RAND has been asked by the 8th Quadrennial Review of Military Compensation (QRMC) to learn about your attitudes towards military compensation. The 8th QRMC is a Department of Defense study group of about 35 people. Every four years, as required by law, the President charters such a group to spend a year and a half studying some aspect of military compensation. The ongoing QRMC is taking a strategic look at military human resource management to propose a system that will meet the needs of the services in the twenty-first century. We need your help to achieve this important task. Specifically, we need your honest thoughts and feelings about five broad topics: (a) what military culture is like, (b) what behaviors are desired and rewarded by the military, (c) what rewards are effective, (d) what is fair and unfair about military pay and benefits, and (e) how people in the military should be compensated, organized, and managed. To achieve these goals, we ask that you complete two short surveys and participate in a focus-group discussion.

Your participation in the surveys and discussion is entirely voluntary. If you do not wish to participate, you may leave. Also, please feel free to not answer any questions that you are uncomfortable with. We will keep everything you say confidential. We will not keep a list of names. We will take notes during our discussion, but we will not insert your names into the notes. The notes will not be shown to anyone outside of RAND and the QRMC. Our report will include only a general description of the people who participated and a summary of the discussions from which the identity and views of the participants cannot be inferred. One last thing: when you leave today, we ask that you not discuss who said what during this focus group in order to protect the privacy of everyone's views.

> Q1: Before we get started, do you have any questions about us, RAND, or the purpose of this discussion or project?

We'll start with a 15-minute survey, then begin an hour-long discussion, and finish with another 15-minute survey. We'll try to finish everything by ____.

Pass out first survey.

Please begin the first survey.

Exhibit A.2

PROTOCOL FOR FOCUS GROUPS

I. Military Culture

In this section we are interested in your perceptions of military culture. First we'd like you to focus on what each of the four services is like.

    Q1: What makes each service distinct?
- P: Why did you choose the service that you did?
- P: Why did you *not* choose the other services?
- P: When you think of the Army, what word or phrase comes to mind?
- P: When you think of the Navy, what word or phrase comes to mind?
- P: When you think of the Air Force, what word or phrase comes to mind?
- P: When you think of the Marines, what word or phrase comes to mind?

You've told us what each service is like; now we'd like to know what the military is like.

    Q2: What sets people in the military apart from people not in the military?
- P: When you think of the military, what word or phrase comes to mind?

We'd like you to think about just your base for the next question.

    Q3: Tell us about some of the different groups at your base.
- P: Who hangs out with whom?
  - –at work
  - –socially, outside of work (by officer vs. enlisted? by occupation?)
- P: What's distinctive about each group?
- P: What word or phrase comes to mind when you think of each group?
- P: Officer vs. Enlisted
- P: Accession source (Officers: ROTC, Academy, or OTS/OCS; Enlisted: recruitment)
- P: Different occupational groups [expand and tailor to each service]
- P: Junior vs. Senior ranks
- P: Active vs. Reserve
- P: Military vs. civilian

    Q4: Are relations between these groups generally positive, negative, neutral, or non-existent?

II. Desired and Rewarded Behaviors

We'd like to ask you a few questions about what people in your rank and occupation are taught is desirable and undesirable behavior.

    Q1: What behaviors do your direct supervisors and commanding officers consider desirable?
- P: Conformity, following orders/protocol
- P: Taking initiative, acting like a leader
- P: Innovative/creative thinking (finding new solutions to old problems)
- P: Helping self versus others
- P: Acting independently (like an individual) versus interdependently (like a group member)
- P: Individual vs. group performance

Q2: What behaviors do your direct supervisors and commanding officers consider undesirable?

Q3: Do you ever get mixed messages about what is expected and valued?

Now, we'd like to ask you about the behaviors that are rewarded and not rewarded.

Q4: What are some behaviors that are rewarded?

Q5: What are some behaviors that are punished?

Now, we would like to ask you about behaviors that are desirable but don't get rewarded and behaviors that are not desirable but do get rewarded.

Q6: What are some *undesirable* behaviors that are rewarded?
    P: How are these undesirable behaviors rewarded?
    P: Why are these undesirable behaviors rewarded?

Q7: What are some desirable behaviors that are *not* rewarded?
    P: Why are these desirable behaviors not rewarded?

Q8: In general, what characteristics make an officer [enlisted] in your unit a good one?

Q9: In general, what characteristics make an officer [enlisted] in your unit promotable?

Q10: Can you think of some good officers [enlisteds] who were passed over for promotion?
    P: Why?

Q11: Can you think of some not so good officers [enlisteds] who received promotions?
    P: Why?

III. Effective Rewards and Punishments

Q1: What are some ways that behaviors get rewarded?
    P: Pay raises
    P: Benefit increases
    P: Promotions
    P: Bonuses
    P: Awards, ribbons, medals
    P: Desirable assignments, task/job duty changes
    P: Praise/recognition from peers, superiors (verbal or written)
    P: School selections
    P: Days off
    P: Perks (parking passes)
    P: Intrinsic rewards (pride, feeling of accomplishment)

Q2: Describe a reward used in the military that makes you work harder.

Q3: Describe a reward not used in the military that would make you work harder.

Q4: Describe a reward used in the military that does not make you work harder.

Q5: What are some ways that behaviors get punished?
    P: Pay cuts or no pay raises
    P: Benefit reductions
    P: Demotion/reduction in grade
    P: Redlining promotion
    P: Difficult, boring, or otherwise undesirable assignments

P: Criticism from peers, superiors
P: Harassment from the system (investigations, threats, letters to the commander)
P: Judicial punishment
P: Letters of reprimand
P: Intrinsic punishment (guilt, regret, shame, embarrassment)

Q6: Describe a punishment used in the military that makes you improve your behavior.

Q7: Describe a punishment used in the military that has no effect on your behavior.

IV. Fairness and Unfairness in Military Compensation

Now I would like to ask you a few questions about your pay, the promotion system, and retirement benefits.

Q1: How should pay be distributed to be fair?
P: Equally? That is, should everyone in the military be paid the same amount?
P: Differently? If so, on which characteristics should pay be differentiated?
P: YOS
P: Rank
P: Occupation
P: Performance
P: Effort
P: Education, knowledge
P: Marital status, number of dependents
P: Officer vs. Enlisted
P: Service

Q2: Should the pay and benefits of military personnel be less than, equal to, or more than the pay and benefits of equivalent civilians? Why?
P: What is an equivalent civilian? Same job, same education and skills, or_____?

Q3: What do you like/dislike about the way promotion decisions are made?
P: Process is clear/mysterious
P: Process is objective/subjective
P: Process considers right/wrong factors
P: Process picks the right/wrong people

Q4: What do you like/dislike about the current retirement system?
P: Vesting after 20 YOS
P: Level of benefits

V. Attitudes Toward Human Resource Management Techniques

Finally, we would like to ask you some questions about ideas to improve the effectiveness of military units.

Q1: First, which of the following would help your unit the most: better individual performance or better teamwork and cooperation among people in the unit?

Q2: What would you think about part of pay being based on individual performance? Could it be measured fairly?

Q3: What would you think about part of pay being based on group performance? This type of pay is referred to as *gainsharing*, whereby everybody in the organization gets a monetary award if the unit keeps its operating costs below a certain level, or *goalsharing*, whereby everybody gets a monetary award if certain organizational goals are met (e.g., a certain in-commission rate for vehicles or some combat qualification level). Do you think a system like this would be helpful for military organizations?

Q4: What would you think about part of your pay being based on individual human capital development, by which we mean anything that can better prepare you to do current or future jobs? Types of human capital development include training, education, physical fitness, marksmanship, and experience in critical jobs in your specialty.

Q5: Would a 360-degree evaluation system improve the performance of your unit?

Q6: Do you think the services should relax lateral-entry restrictions, i.e., allow individuals with needed skills, qualifications, and experience to enter military service at middle or senior officer and NCO grades? Do you think retirement benefits should be vested before 20 years of service, to make it easier for people to migrate out of the military at mid-career points?

Q7: Do you think there is room in the military for self-managed teams, i.e., teams that manage their own tasks and interpersonal processes, without first-line supervisors?

Exhibit A.3

OPEN-ENDED SURVEY (COMPLETED BEFORE FOCUS-GROUP DISCUSSIONS)

I. General

Please provide the following basic information about yourself.

1. Service: _____
2. Rank/rate: _____
3. MOS, AFSC, or designator/NEC: Code: _____, Occupational Title: _____
4. Date entered active duty (month and year): _____
5. Age: _____
6. Race: _____
7. Sex: ___Male, ___Female
8. Education (check highest level earned): ___some high school, ___high school diploma, ___some college, ___associate's degree, ___bachelor's degree, ___master's degree, ___doctorate degree
9. Accession Source: ___Academy, ___ROTC, ___OTS/OCS, ___Recruit
10. Marital Status: ___Married, ___ Separated, ___ Divorced, ___Single
11. If married, is your spouse: ___in the military, ___not in the military and employed, ___not in the military and not employed.
12. Number of dependents (including spouse): ____
13. Month and year of your most recent promotion/advancement (date you pinned or sewed on your current grade): _____
14. (Officers only) Have you ever been promoted below the zone? ___Yes, ___No
15. (Officers only) Have you ever been promoted above the zone? ___Yes, ___No
16. About how many hours do you work in a typical week? _____
17. How many of the following personal rewards have you received?
    -achievement medals: _____
    -commendation medals: _____
    -meritorious service medals: _____
    -legion of merit medals: _____

18. Compared to your peers of the same rank and occupation, how would you rate your overall performance?

```
1--------- ------2-------- -------3------------- ---4-------- -------5----- -------6----- ------7
much worse    moderately    slightly     about equal    slightly     moderately  much better
than my peers   worse         worse        to my peers    better        better      than my peers
```

19. Compared to your peers of the same rank and occupation, how do you think your direct supervisors and commanding officers would rate your overall performance?

```
1--------- ------2-------- -------3------------- ---4-------- -------5----- -------6----- ------7
much worse    moderately    slightly     about equal    slightly     moderately  much better
than my peers   worse         worse        to my peers    better        better      than my peers
```

For the next few sections, we would like you to write down a few phrases, whatever comes to mind quickly. There's no need to think and write too much about each question.

II. Desired and Undesired Behaviors

    1. Write some specific behaviors your direct supervisors and commanding officers **want to see** more often from you at work.

    2. Write some specific behaviors your direct supervisors and commanding officers **want to see** less often from you at work.

III. Rewarded and Punished Behaviors

    1. Write some specific behaviors your direct supervisors and commanding officers **rewarded you** for doing. (If you cannot think of anything you have been rewarded for doing, write some specific behaviors your direct supervisors and commanding officers rewarded someone else for doing.)

    2. Write some specific behaviors your direct supervisors and commanding officers criticized or punished you for doing. (If you can't think of anything you've been criticized or punished for doing, write some specific behaviors your direct supervisors and commanding officers criticized or **punished someone** else for doing.)

IV. Kinds of Rewards and Punishments

    1. How do your direct supervisors and commanding officers reward behavior **they want from you?**

    2. How do your direct supervisors and commanding officers punish behavior **they do not want from you?** (Be sure to include mild forms of punishment such as negative feedback.)

    3. How else (besides rewarding and punishing you) do your direct supervisors **and commanding** officers encourage you to do what they want from you?

V. Attitudes Toward Military Compensation:

    1. What do you like most about military pay and benefits?

    2. What do you dislike most about military pay and benefits?

    3. Do you think basic pay should depend on rank? ____ Yes, ____ No
    4. Do you think basic pay should depend on years of service? ____ Yes, ____ No
    5. What other factors should basic pay depend on?

Exhibit A.4

CLOSE-ENDED SURVEY (COMPLETED AFTER FOCUS-GROUP DISCUSSIONS)

I. Desired Behaviors:

Listed below are many behaviors. In the first column, we would like you to indicate how **desirable** (or undesirable) the behavior is in the eyes of your direct supervisors and commanding officers. For example, how desirable would your superiors judge your wearing a clean and neat uniform? In the second column, we would like you to indicate how **rewarded** (or punished) the behavior is by the military. For example, how much would the military reward you for wearing a neat and clean uniform?

Use the scale below to rate how **desirable** each behavior is in the eyes of your superiors.

```
1----------------2----------------3------------4------------5---------------6----------------7
extremely   moderately   slightly    neutral   slightly    moderately   extremely
undesirable undesirable  undesirable           desirable   desirable    desirable
```

Use the scale below to rate how **rewarded** each behavior is by the military.

```
1----------------2----------------3------------4------------5---------------6----------------7
punished    punished     punished    neither    rewarded    rewarded     rewarded
severely    moderately   mildly      punished   a little    moderately   a lot
                                     nor rewarded
```

How Desirable?	How Rewarded?	Behavior to be rated
_____	_____	Obeying orders of my superiors.
_____	_____	Respecting authority.
_____	_____	Conforming to the norms and values of the military.
_____	_____	Working cooperatively with my peers.
_____	_____	Competing with my peers.
_____	_____	Relying on my peers when I need help (being dependent).
_____	_____	Relying only on myself for help (being independent).
_____	_____	Helping my peers to achieve their goals.
_____	_____	Helping my unit to achieve its goals.
_____	_____	Helping myself to achieve my goals.
_____	_____	Putting the interests of others before my own.
_____	_____	Acting like an individual first and a member of my unit second.
_____	_____	Taking initiative; acting like a leader.
_____	_____	Thinking creatively (finding new solutions to old problems).
_____	_____	Exerting more effort at my job.
_____	_____	Having a good attitude at work.
_____	_____	Spending more time at my job.
_____	_____	Being more productive at my job.
_____	_____	Making fewer mistakes at my job.
_____	_____	Saving time (doing my job in less time).
_____	_____	Saving money or cost-cutting (doing my job with less money).
_____	_____	Maintaining physical fitness.
_____	_____	Marrying.
_____	_____	Having children.
_____	_____	Socializing with peers after work.
_____	_____	Being intellectual.

II. Effectiveness of Rewards

The military uses many kinds of rewards and punishments. We are interested in how effective they are at motivating desired behavior.

1. Based on your experience in the military, how much has each of the following factors influenced how hard and how well you have worked. Use the scale below.

```
1-------------------------------2-------------------------------3-----------------------------4
had no effect          had a small effect        had a moderate effect        had a large effect
on your work           on your work              on your work                 on your work
```

_____ Your pay.
_____ Your benefits.
_____ Amount of praise you received for desirable behavior.
_____ Amount of criticism you received for undesirable behavior.
_____ Your working conditions.
_____ How much you liked working and socializing with your peers.
_____ How much you respected your direct supervisors and commanding officers.
_____ Your job assignment.
_____ Your base location.
_____ How fairly you were treated in promotion decisions.
_____ Whether you felt like a valued and respected member of your unit.
_____ How much choice you were given about your job assignment.
_____ How much choice you were given about where you were stationed and for how long.

2. Use the scale below to rate how each reward or punishment would affect the quality and quantity of your work. For example, if receiving $100/month more in basic pay would make the quality and quantity of your work "slightly better," then write a "5" in the space.

```
1----------------2---------------3---------------4-----------5-----------6--------------7
much          moderately      slightly       no effect   slightly   moderately      much
worse         worse           worse                      better     better          better
```

_____ Receiving $100/month more in basic pay.
_____ Receiving $200/month more in basic pay.
_____ Receiving $100/month less in basic pay.
_____ Receiving $200/month less in basic pay.
_____ Compared to being stationed at an average location, being stationed at a very desirable location.
_____ Compared to being stationed at an average location, being stationed at a very undesirable location.
_____ Compared to being assigned to an average job, being assigned to your favorite realistic job.
_____ Compared to being assigned to an average job, being assigned to your worst realistic job.
_____ Receiving more praise for desirable behavior from your superiors.
_____ Receiving less praise for desirable behavior from your superiors.
_____ Receiving more criticism for undesirable behavior from your superiors.
_____ Receiving less criticism for undesirable behavior from your superiors.
_____ Receiving a promotion you feel you deserve.
_____ Not receiving a promotion you feel you deserve.
_____ Being given more choice about your job assignment.
_____ Being given more choice about where you are stationed and for how long.

98 Differentiation in Military Human Resource Management

III. Attitudes Toward Military Compensation:

1. Which statement best characterizes your thinking about your level of pay and benefits. (Check one and then explain your response below.)

_____ Overall, I am very unsatisfied. I think I receive way too little pay and way too few benefits.
_____ Overall, I am unsatisfied. I think I receive too little pay and too few benefits.
_____ Overall, I am satisfied. I think I receive about the right amount of pay and benefits.
_____ Overall, I am very satisfied. I think I receive a generous amount of pay and benefits.

Explain:

2. Compared to my peers in the military, I think my level of pay and benefits is (check one and explain)
_____ way too little
_____ too little
_____ about right
_____ too much
_____ way too much

Explain:

3. Compared to my peers outside the military, I think my level of pay and benefits is (check one and explain)
_____ way too little
_____ too little
_____ about right
_____ too much
_____ way too much

Explain:

4. Which factors do affect and should affect total compensation (includes all forms of compensation such as basic pay, special pay, bonuses, housing, food, healthcare, and retirement)? Circle Yes or No for each question. For example, if you think years of service does affect your total compensation, circle Yes in the space under "Does affect?" If you think years of service should affect your total compensation, circle Yes in the space under "Should affect?" Please add any comments you think would help us to understand what does and should affect total compensation.

Does affect?	**Should affect?**	**Factor:**
Yes/No	Yes/No	years of service
Yes/No	Yes/No	rank/rate
Yes/No	Yes/No	individual performance (how well you do your job)
Yes/No	Yes/No	group performance (how well your unit does its job)
Yes/No	Yes/No	knowledge (as measured by tests)
Yes/No	Yes/No	education
Yes/No	Yes/No	effort (how hard you work)
Yes/No	Yes/No	physical fitness
Yes/No	Yes/No	accession source
Yes/No	Yes/No	Officer vs. Enlisted
Yes/No	Yes/No	occupation
Yes/No	Yes/No	marital status

Yes/No	Yes/No	number of dependents
Yes/No	Yes/No	cost of living in your area
Yes/No	Yes/No	being at war vs. at peace
Yes/No	Yes/No	service

Comments:

IV. Attitudes Toward Human Resource Management Techniques

People can be compensated, managed, and organized in many different ways. We would like to know your attitude toward some different ways of compensating, managing and organizing people. Use the following scale to rate your attitude toward each technique.

```
1---------------2-----------------3-----------------4---------------5-------------6--------------7
extremely    moderately    mildly      indifferent    mildly    moderately    extremely
unappealing  unappealing  unappealing                 appealing  appealing    appealing
```

_____ Individual performance as a basis for promotion
_____ Team or organizational performance as a basis for promotion
_____ Pay for individual performance
_____ Pay for team or organizational performance (gainsharing or goalsharing)
_____ Pay for human capital development (training, education, experience, physical fitness, etc.)
_____ 360-degree evaluation system (evaluation by peers and subordinates, as well as by supervisors)
_____ Lateral entries (direct appointment of qualified individuals at middle or senior grades)
_____ Earlier vesting of retirement benefits (being eligible for retirement benefits before 20 YOS)
_____ Self-managed teams (teams that manage their own task and interpersonal processes, without first-line supervisors)

Appendix B
LINKING HRM TO BEHAVIOR—THEORY, EVIDENCE, AND IMPLICATIONS

As we contemplated new and existing departures from uniformity in military HRM systems, we identified three broad areas of interest: how rewards could be more effectively differentiated to motivate desired behavior, what questions of distributional fairness the issue of differentiation would bring to the forefront, and the higher costs that greater differentiation might make in an HRM system.

Bodies of research relevant to these areas of interest are reported in several behavioral science and economics literatures. Three theories in particular, drawn from that research, can be readily applied to the case of military HRM:

- *Expectancy* theory (Vroom, 1964; Porter and Lawler, 1968) relates behaviors to rewards in an organizational context, explicitly recognizing the role of organizations (and their HRM systems) in mediating rewards.

- *Social justice* theory (Adams, 1963; Greenberg, 1987; Sheppard, Lewicki, and Minton, 1992) describes how people develop a sense of fairness about the distribution of rewards and how this sense of fairness affects their behavior.

- *Transaction cost economics* theory (Williamson, 1975) describes how parties to a contract, including employers and employees in an employment contract, find the least-costly ways of relating to each other under varying conditions.

In this appendix, we briefly describe each of these theories, characterize their empirical support, and discuss some of the broad implications we see in them. We also provide a wider list of theories and bodies of research that have some bearing on the issues we examine in the main body of the report.

EXPECTANCY THEORY

Expectancy theory holds that motivation can be enhanced by strengthening two expectancies through which rewards are associated with performance and by increasing the valence (strength, or attractiveness) of rewards.

Strengthening the Two Expectancies

The first expectancy—that individual or group behaviors will result in an outcome of value to the organization—is largely a function of organizational structure and job design. It is strengthened if work is organized so that all jobs are perceived to be worth doing and so that workers can see how their efforts contribute to a productive whole. This element of expectancy theory has important implications for how functional managers and local supervisors arrange work processes, but it also has corresponding implications for personnel managers.

It suggests broadening jobs and job classifications as much as possible so that each worker (for separable tasks—those in which an individual can produce a meaningful output) or group of workers (for inseparable tasks—those requiring more than one person to produce a meaningful output) can more readily identify their contributions to measurable outputs. It also suggests that work processes should be arranged to minimize the influence of environment or other forces beyond workers' control on outcomes used as a basis of reward. Finally, the linkage between effort/performance and outcomes can be a function of workers' perceptions that they have adequate human capital and organizational resources to do their jobs.

The second expectancy—that the organization will observe individual or group outcomes and reward them—is strengthened by forging stronger links between desired outcomes and rewards. Perhaps the simplest way to strengthen the link is to decrease the time interval between outcomes and rewards. For example, in a military compensation system that typically spaces merit-based promotions three to six years apart,[1] the implication is that more-frequent intragrade merit pay increases or bonuses could be useful.

Another way to strengthen the link is to reduce the number of outcomes that co-determine rewards. In the military system, for example, the implication is that effort/performance and human capital development might be more effectively motivated if they are separately, rather than jointly, rewarded. The theory argues for selecting the appropriate individual or group basis for rewards—individual rewards for separable tasks and group rewards for inseparable tasks. Finally, this linkage can be improved by increasing the accuracy of outcome measures: improving the organization's observation of worker contributions. For example, most rewards in the current military HRM system are based on subjective assessments of performance, which underscores the need for continued efforts to refine formal evaluation systems.

---

[1] Promotions up to grades E-4 and O-3 are generally based on time in grade, with very limited consideration of performance or other indicators of merit. For promotions from E-4 to E-9 and from O-3 to O-6, average number of years between promotions were as follows (computed from data on average time in service at promotion, found in *Uniformed Services Almanac*, 1995, p. 234):

	Army	Navy	Air Force	Marines
Officer	5.5	5.3	5.4	5.6
Enlisted	3.8	3.5	3.7	3.8

Increasing the Valence of Reward

The theory also predicts that motivation can be enhanced by increasing the valence of rewards. The Seventh QRMC's recommended restructuring of military pay tables to increase the returns to promotion and decrease the returns to longevity is seen as a means of increasing the valence of rewards for performance (Seventh QRMC, 1992). This restructuring would also, of course, increase the valence of rewards for human capital development and other behaviors that currently influence promotion selection.

Empirical Support for the Theory

Porter, Lawler, and Hackman (1975, p. 57) assessed empirical support for expectancy theory as follows:

> Most researchers who have applied expectancy theory to work organizations have asked employees to report on their expectancies and valences. Strict behaviorists would reject this approach and predict behavior by a careful analysis of the situation and the past behavior of the people involved. Studies have shown, however, that people's reports of their expectancies and valences *can* predict later behavior, although not all studies have found that valence measures are useful.... Thus, while there are some technical difficulties associated with exact formulations of expectancy theory, there seems to be widespread research support for the utility of a general expectancy-theory approach to understanding the behavior of people in organizations.

Dresang (1991) cites Vroom's (1964) work as the major body of support for this theory. He also cites a number of other scholars who have confirmed both parts of the expectancy-theory sequence and the theory as a whole: Atkinson (1964); Ivancevich, Szilagyi, and Wallace (1977); and Campbell, Dunnette, Lawler, and Weick (1970). Dresang (1991, pp. 230–232) finds that a combination of expectancy theory and equity theory (see "Social Justice Theory" below), with supporting research, provides a basis for policies and procedures having a balance of emphasis on performance and fairness that is appropriate for the public sector.

Implications of Theory for Military HRM

Compensation consultants refer to strengthening the links between effort/ performance and organizational outcomes and between organizational outcomes and rewards as improving the *line of sight* (Lawler, 1990). This is the primary focus of most performance-related compensation enhancements. Military HRM systems present rich possibilities for improving the line of sight, some of which are discussed in Chapter Three. However, potential line-of-sight adjustments must be carefully analyzed to ensure that influences on equity, group cohesion, and other important considerations are anticipated and given appropriate weight.

Social Justice Theory

Adams (1963) describes equity theory as a special case of Festinger's (1957) theory of cognitive dissonance. The theory holds that individuals judge the fairness of a social exchange, such as employment, by comparing their personal inputs (education, experience, effort, etc.) and personal outcomes (pay, status, etc.) with those of some relevant person or group. People will feel uncomfortable if they perceive that outcomes are not proportional to inputs. In a compensation context, a sense of inequity develops if individuals perceive that they are overexerting or underexerting relative to others receiving the same pay, or if they are overpaid or underpaid relative to others with similar inputs. The theory predicts that those sensing an inequity will take some action to restore balance, such as changing their inputs, attempting to change outcomes, reevaluating inputs or outcomes, or leaving the field.

Sheppard, Lewicki, and Minton (1992) place equity in a larger framework of organizational justice. They see judgments about fairness made on the basis of two principles: balance (comparability with similar actions in similar situations), and correctness (consistency, accuracy, clarity, procedural thoroughness, and compatibility with the morals and values of the times). They see justice evaluated at three levels: the outcome of an action, which many writers refer to as *distributive justice*; the procedure that generates and implements the outcome, which many writers refer to as *procedural justice*; and the system within which the outcome and the procedure are embedded, referred to as *systemic justice*. To complete their framework, they see three goals that are pursued in the application of justice: *performance effectiveness* (maximizing the quantity and quality of output produced by individuals or groups), *sense of community* (a sense of membership in, and identity with, some social entity), and *individual dignity and humaneness* (a sense of well-being, individual identity, and personal worth).

From this theoretical framework, Sheppard, Lewicki, and Minton develop standards of fairness using both principles, across all levels, and with respect to each goal. Their three standards for balance at the outcome level seem especially relevant to compensation issues. Application of a performance goal results in a standard of *equity*—rewards consistent with the quantity and quality of results produced. A community goal suggests a standard of *equality*—equal outcome, regardless of performance or effort. Finally, a concern for human dignity and humaneness leads to a standard of *need*—meeting requirements appropriate to one's station in life.

Empirical Support for the Theory

Sheppard, Lewicki, and Minton (1992, pp. 102–103) cite empirical findings that support three effects of persistent justice:

> First, justice usually produces immediate and direct consequences. Thus, equitable pay improves individual performance, equal treatment raises group spirit, voice creates greater commitment to a decision, and access creates a loyal ally.

> Second, persistent justice engenders positive affect toward the agent. Thus, distributive justice tends to make employees feel good about their jobs, procedural

> justice creates loyalty toward management, and systemic justice creates loyalty toward the firm and the firm's objectives. . . .
>
> The third and perhaps most important impact of continued justice is that it protects the agent from active, negative, and group responses to the occasional injustice. We are all imperfect; an organizational history of just behavior creates a tendency to overlook infrequent injustices. . . .

Cowherd and Levine (1992) assess the extent to which distributive justice theory has been tested in organizational contexts:

> There has been extensive research on the effects of pay equity on work attitudes and behavior, such as pay and job satisfaction (e.g., Oldham et al., 1986), absenteeism (e.g., Dittrich and Carrell, 1979), sickness and accident compensation costs (e.g., Sashkin and Williams, 1990), turnover (e.g., Telly, French, and Scott, 1971), and work performance (e.g., Pritchard, Dunnette, and Jorgenson, 1972; Summers and Hendrix, 1991).

Lind and Tyler (1988) similarly assess empirical support for procedural justice theory. They cite an extensive list of laboratory and field studies supporting four of the most important procedural justice effects: (1) the enhancement of procedural justice judgments when those affected by the procedure are granted voice or process control, (2) the enhancement of distributive justice judgments and satisfaction with outcomes by procedural justice judgments, (3) the enhancement of attitudes toward authorities by judgments of procedural justice, and (4) the instigation of various salutary behavioral effects by judgments of procedural justice. Although they find a few disconfirming studies, their conclusion is that procedural justice effects are robust across methodologies.

Implications of the Theory for Military HRM

This body of theory has conflicting implications for differentiation in military personnel management and compensation. An equity standard implies greater differentiation of rewards as a path toward improved individual performance, an element of the broader set of behaviors we have defined within effort/performance. It also supports differentiation based on any human capital development for which the individual bears the input costs. An equality standard, however, implies less differentiation, supporting community goals (a feeling that "We are all in this together"), which may play an important role in intrinsic rewards and in elements of effort/performance, such as teamwork and collaboration.

There is a perception that movement from equality toward equity, or from lesser to greater differentiation, is needed in military systems (Seventh QRMC, 1992; Asch and Warner, 1994). Performance goals favor such movement, but we see a need to proceed cautiously.

First, we would consider the net effects on unit productivity and readiness of marginal improvements in individual performance, coupled with marginal decrements in community effects. In circumstances where individual-performance improvements would have the greater weight, more differentiation would be indicated.

Where community effects would have greater weight, less differentiation would be indicated.[2]

Second, we would predicate greater differentiation on effective procedural justice. If differences in performance cannot be reliably measured through subjective appraisals or through other, more objective approaches, greater differentiation of rewards could easily do more harm than good.

TRANSACTION COST ECONOMICS

The theory of transaction cost economics (Williamson, 1975) examines the arrangements through which parties interact in contractual relationships. Its primary focus is on external contracts between organizations, but it has also been applied to internal contracts between organizations and their employees.

Transaction costs are associated with the overhead functions in a contracting process—search, negotiation, monitoring, and enforcement.[3] The first two functions occur ex ante, i.e., before a contract is established. The third and fourth functions occur ex post, i.e., during the period of performance after a contract is established. A body of theory helps to select among varying contract characteristics, or *governance structures*, depending on the contracting environment, in order to minimize these transaction costs. The theory is particularly salient on three characteristics of contracts—frequency, completeness, and duration.

Frequency

The theory recognizes that the more people engage in productive relationships (frequency), the more they learn to trust each other. Trust can substitute for information about whether parties to an agreement are completing their respective tasks, which means greater use of more-informal commitments to ensure performance results. Less-frequent transactions usually generate less trust among contractual parties; therefore, more-formal commitments generating greater organizational control of resources used in these transactions may be the best choice.

Completeness

As a general rule, the theory predicts that transaction costs increase as agreements become more complete and as the environment becomes more complex. However, benefits are associated with a complete agreement—e.g., less ex post haggling over

---

[2] Lazear (1995) captures this relationship in a tournament-theory framework.

[3] *Search* refers to the process of finding a suitable other party for a contract. In a labor market transaction, it refers to recruiting on the part of the organization and job hunting on the part of the individual. *Negotiation* refers to the process of arriving at terms for the contract. *Monitoring* is observation by each party of the others' behavior. *Enforcement* is action taken by a party to ensure that the other party meets the terms of the agreement.

payoffs—that increase as the agreement becomes more complete.[4] These benefits also tend to rise as a function of complexity. *Complexity* is related to asset specificity (the extent to which one party has assets that the other cannot easily obtain elsewhere and therefore has the upper hand in haggling). In employment relationships, asset specificity is generally considered to be a function of the degree to which workers develop firm-specific human capital.[5]

Complexity is also related to the separability of tasks. Environments in which outcomes rely on group performance are more complex than those in which outcomes rely on individual performance. For any employment agreement, organizations and workers would seek an optimal degree of completeness where marginal costs equal marginal benefits.

Duration

Transaction costs and benefits also increase as the duration of an agreement increases and, again, as a function of complexity in the environment. For example, for longer agreements, the parties might have to consider adjustment of payoff values for inflation or contingencies in which organizational priorities change. Again, organizations and workers seek an optimal duration where marginal costs equal marginal benefits.

Empirical Support for the Theory

Masten (1993) assessed the state of empirical support for the theory. He found that empirical research to date supports the view that transaction-cost considerations influence organizational choices. This research tends to test positive, or descriptive, propositions (i.e., that managers tend to choose governance structures in the way that the theory predicts). He cautions, however, that a theory that is a good predictor of how managers make organizational choices may not be a good predictor of organizational performance. He finds that strategic-management literature has provided few incontrovertible insights into the value to managers of particular theories of organizational choice, and attributes this paucity to both conceptual and practical difficulties in relating organizational form to performance. To support the normative validity of the theory, he sees a need for a more rigorous body of strategy research.

Implications of Theory for Military HRM

Transaction cost economics has implications for one of the salient characteristics of military HRM—its strongly internalized labor market. Transaction cost economics theory sees an internal labor market as a somewhat incomplete but long-duration governance structure. Although costs may be higher in an internal labor market, the

---

[4]Haggling over payoffs is a form of what transaction cost economics theorists call *opportunistic behavior*, defined as that by which one party seeks to alter the balance of benefits at the expense of the other party.

[5]We use the term *firm-specific*, which has a private-sector connotation, to conform to usage in the transaction cost economics literature. A more generally applicable term might be *organization-specific*.

theory predicts that it is the best arrangement for a complex environment because benefits also rise. (The arrangement makes marginal costs equal marginal benefits.) The theory suggests that, in the absence of high asset specificity, the marginal costs incurred in internal labor markets may exceed marginal benefits. In skills or functions that do not demand high levels of firm-specific human capital, transaction costs can be reduced by moving to a more externalized labor market. In practical terms, the theory suggests that military HRM systems might be usefully differentiated across units or functions by varying the barriers to lateral entry and exit.

Transaction cost economics also provides some insight into the structuring of differentiated rewards. When rewards are differentiated, managers of line organizations and HRM systems must establish rules for distributing the rewards. Both costs and benefits rise as rules become more complete. In less-complex environments, the theory predicts that relatively complete rules will equate marginal costs and benefits. The ex ante cost of formulating the rules is not high, and benefits in terms of reduced ex post monitoring costs and haggling can be realized. In more-complex environments, the theory predicts that rules will be relatively incomplete. The ex ante costs of trying to anticipate all contingencies become higher than the ex post benefits.

In current military HRM, existing governance structures impose some transaction costs on central HRM activities and some on line managers. The theory is indifferent with respect to where these costs are imposed. It predicts the behavior that will minimize the sum of them. Central HRM functions—such as recruiting and accession-processing functions, training and education activities, headquarters personnel staffs, and personnel centers—are highly visible manifestations of HRM transaction costs. The roles line managers play in motivating, evaluating, counseling, disciplining, and otherwise attending to the behaviors of subordinates represent unseen and unmeasured HRM costs. The theory seems to remind central HRM managers, who design most features of the governance structures that impose costs on line managers, that they, too, should consider the sum of transaction costs.

OTHER THEORIES AND RESEARCH

The landscape of compensation and organizational theory is rich and varied. Although we chose a few from among the many theories to help order our thinking on differentiation of military HRM, others might also have served our purpose. In the following paragraphs, we briefly sketch our rationales for the choices we made.[6]

Compensation Theory

In choosing from among compensation theories (see Table B.1), we leaned toward the behavioral sciences rather than the labor economics literature. Labor economics theory is more nuanced in explaining how and why labor markets shape reward sys-

---

[6]We would not insist that the theoretical frameworks we preferred are necessarily the best available for examining our topic. Other researchers might have preferred other frameworks and might have reached conclusions of equal or greater validity. Our purpose here is simply to show that our choices were rational, even if boundedly so, rather than arbitrary.

tems, but it uses a summary construct—maximizing utility—to explain individual behavior. Behavioral science is more nuanced in explaining how and why individuals react as they do to *differentiated* rewards and offered more help in exploring the many intervening variables linking rewards to behaviors. Another advantage offered by the behavioral science literature is its concreteness. Whereas labor economics theories tend to be more abstract, expressing important relationships mathematically, the behavioral science literature tends to describe behavior in the terms more likely to be used by the line and human resource managers responsible for formulating HRM policies.

Among the behavior science theories, we chose expectancy theory over closely related drive theories and social justice over closely related relative-deprivation theory for similar reasons: richness in the bodies of contemporary supporting research and concreteness related to differentiation in HRM systems.

Table B.1

Compensation Theories

Theory	Focus	References
Labor Economics		
Agency	Contingency pay as a sharing of outputs between principals and agents	Ross, 1973; Holmstrom, 1979
Tournament	Compensation as a function of rank rather than output	Lazear and Rosen, 1981; Rosen, 1986
Human capital	Economic returns to investments in education, training, health, and other individual characteristics	Becker, 1975
Segmented markets	The occurrence of wage differentials that do not correspond to skill differentials	Kerr, 1954; Taubman and Wachter, 1986
Efficiency wage	Why some firms offer wages above the market even in the face of unemployment	Stiglitz, 1976; Yellen, 1984; Akerlof, 1984
Equalizing differences	Differences in working conditions give rise to differences in wages	Rosen, 1986
Behavioral Sciences		
Drive	Behavior as motivated by a need to relieve physiological or psychological deprivations	Maslow, 1943; McGregor, 1960; McClelland, 1961; Herzberg 1968
Expectancy	Expectancies and valences as linkages between compensation and effort motivation	Vroom, 1964; Porter and Lawler, 1968
Social justice	How employees evaluate and react to allocative decisions	Adams, 1963; Greenberg, 1987
Relative deprivation	Why some clearly disadvantaged people do not find their positions unjust	Crosby et al., 1986; Martin and Murray, 1983

Organizational Theory

In examining the body of organization theory (see Table B.2), we found much work dedicated to explaining how and why organizational structures, including HRM structures, have evolved. Our specific interest was in exploring the intervening variables linking behavior to these structures. We felt that transaction cost economics, with its focus on governance structures, provided the richest insights into these linkages (i.e., into how behavior might be affected by differentiation in HRM structures).

Table B.2

Organizational Theories

Theory	Focus	References
Classical and Neo-Classical Schools		
Scientific management	The optimal configuration of formal organizations, seen as closed, rational systems	Taylor (1911)
Administration	Principles of administration	Fayol (1949)
Bureaucracy	Characteristics of highly efficient organizations	Weber (1946)
Human relations	The role of individual characteristics and behavior	Mayo (1945)
Institutionalization	The process by which organizations develop distinctive structures, capacities, and commitments	Selznick (1948)
Decisionmaking/ bounded rationality	How decisions are made in organizations; hierarchy of ends; satisficing vs. optimal solutions	Simon (1945); March and Simon (1958)
Contemporary Theories		
Contingency	Impact of environmental conditions on organizational design decisions	Lawrence and Lorsch (1967)
Transaction cost economics	The costs of alternative governance structures for the exchange of goods or services	Williamson (1975)
Organizational ecology	Organizational survival and adaptation to environments	Hannan and Freeman (1977); Aldrich (1979); Pfeffer and Salancik (1978)
Marxist	Organizational structures as power systems rather than rational systems	Braverman (1974); Edwards (1979)

NOTE: See Scott (1987) for a more complete taxonomy of organization theory.

Appendix C
ACCOUNTABILITY IN MILITARY ORGANIZATIONS

This appendix examines the issue of accountability in military organizations, looking specifically at patterns of accountability, mechanisms for accountability, and the challenges that military organizations face in accommodating shifting patterns of accountability.

DEFINING *ACCOUNTABILITY*

Accountability may be defined as the process by which senior stakeholders in a relationship observe and enforce the degree to which other parties meet the terms of the relationship. *Senior stakeholders* are those who have the broadest interests, such as the owners or stockholders of private firms or the electorates of government bodies. Senior stakeholders may be thought of as *principals* in the relationship, while the other parties may be thought of as *agents*. Agents use resources supplied by principals to produce some output or outcome of value to the principals. Principals agree to compensate the agents. Principals must find prudent means to ensure that agents are serving their interests.

Accountability cascades through an organization, especially when it is enforced through hierarchical mechanisms. Senior stakeholders employ or elect officeholders, who appoint managers, who oversee workers. At each level, agents are accountable to their superiors and must in turn hold their subordinates accountable.

DIFFERENCES BETWEEN PUBLIC- AND PRIVATE-SECTOR ACCOUNTABILITY

Accountability tends to be more complex in public-sector agencies than in private-sector firms. In the private sector, managers and workers are ultimately accountable to well-defined stakeholders (owners or shareholders) with narrowly defined private interests. Profit, loss, and cost interests predominate and are routinely reported at all levels; survival of the organization and employment at the pleasure of superiors provide ready enforcement mechanisms. With clear interests and strong mechanisms for holding managers and workers to those interests, accountability in the private sector has not drawn significant levels of analytical attention.

In the public sector, accountability is to broad classes of stakeholders with wide, perhaps ill-defined, and even conflicting interests. The ultimate accountability mechanism is the electoral system, which seldom provides clear mandates on specific issues. A system of checks and balances among the branches of government provides another accountability mechanism, but one that tends to leave many conflicting views of the public interest unresolved. At lower levels, a variety of bureaucratic, legal, political, accounting, professional, and other oversight processes serve as accountability mechanisms (Romzek and Dubnick, 1994). These processes seldom reveal success in meeting public-sector objectives as crisply as the financial reporting mechanisms available in the private sector. With diverse interests, difficulty in determining if those interests are met, and less-direct enforcement mechanisms, accountability is difficult to achieve in the public sector and, thus, warrants continuing study and analysis.

PATTERNS OF ACCOUNTABILITY

Figure C.1 illustrates several hierarchical and nonhierarchical patterns of accountability embedded in a systems framework (input, process, output, and outcome). In this model, workers are given inputs in the form of organizational resources, the authority to employ or expend them, rules for using them, and, usually, some discretion in using them.[1] To these organizational inputs, they add their own efforts, decisions, and compliance with rules through processes that yield outputs useful to the organization. Organizational outputs interact with the environment to produce outcomes. Superiors hold subordinates accountable by observing their inputs (effort, decisions, compliance), the outputs they produce, and the resulting outcomes. Accountability may also be enforced by stakeholders who are not direct superiors, through the several nonhierarchical alternatives (discussed in more detail below) indicated in the figure.

The essential elements of a hierarchical, or *bureaucratic*, pattern of accountability are indicated by the small loop of bold black arrows in Figure C.1.[2] In this pattern, accountability is ensured primarily through superiors' strict control of resources, limited grants of authority, firmly established rules, and close observance of subordinates' input to process.

In more-complex and less clearly understood environments (health care, for example), organizations cannot ensure desired outcomes by specifying rules in advance. Greater discretion must be extended to subordinates, who must then be held accountable for outputs and perhaps even outcomes. Superiors may be unable to directly observe outputs and outcomes, may be unqualified to evaluate them, or

---

[1] *Rules*, or *policies*, might specify the purposes for which resources are to be used as well as the manner in which they will be employed or expended.

[2] We use the term *bureaucracy* in the sense intended by Weber (1946), the early-twentieth-century German sociologist and economist, for whom the term denoted a highly efficient form of organization for the administration of a public bureau. In Weber's sense, bureaucracy is characterized by specialization, hierarchy, levels of graded authority, appointment to office based on technical expertise, and management according to well-developed rules.

Accountability in Military Organizations 113

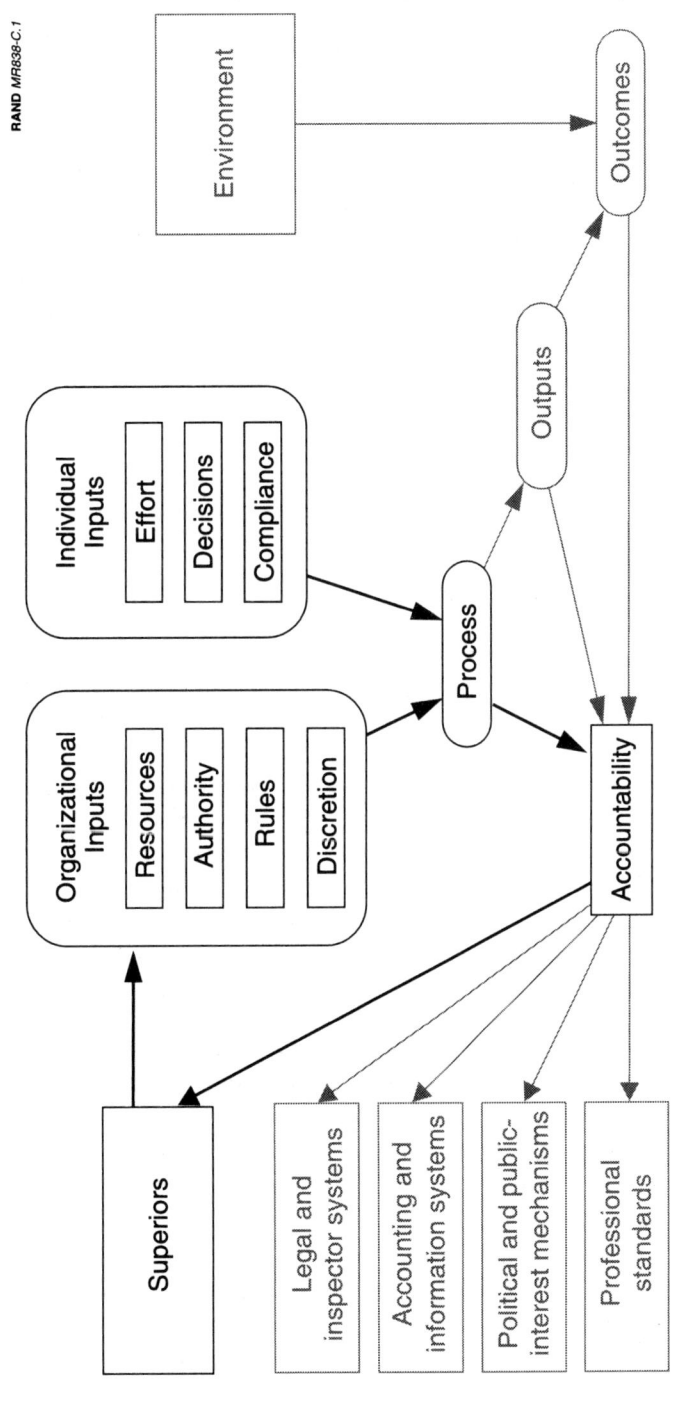

SOURCE: Some elements of the figure are suggested in Romzek and Dubnick (1994).
NOTE: Bolder black elements indicate a hierarchical, or bureaucratic, pattern of accountability; lighter gray elements indicate nonhierarchical patterns.

Figure C.1—Patterns of Accountability

may not have sufficient standing to pass judgment on them. These conditions increase reliance on nonhierarchical alternatives. In Figure C.1, these alternatives are depicted by the wider loop of light gray elements.

Discretion and Accountability in Compensation Systems

Differentiation in compensation generally occurs through the exercise of discretion by decisionmakers at several levels in an organization. In the public sector, broad discretion is exercised by voters in the electoral process and then by elected officials who enact legislation, issue Executive Orders, and make appointments giving shape and direction to various programs. Within the Executive Branch, discretion to further differentiate is exercised at various layers of departmental, agency, and service management, where policies, tables, procedural rules, budgets, and other delegating and controlling mechanisms are developed. At the lowest levels of an organization, discretion is exercised by individuals acting in various line and staff capacities to apply rules and policies to specific cases.

The amount of housing allowance a military member receives provides an example of how discretion exercised at almost every level affects a compensation outcome. The Congress, acting on behalf of the public, establishes a statutory entitlement for military members to be given housing in kind or a cash equivalent. It also authorizes and appropriates funds for military pay and the construction and maintenance of housing. The Congress at times sets the precise amounts of cash equivalents to be paid in lieu of in-kind housing and at other times gives the President some discretion in setting or adjusting amounts, particularly when amounts are varied geographically. The Department of Defense, the services, their major commands, and their installation managers determine local supplies of housing through a budgeting process and may also establish rules for assigning government quarters. A commander decides whether a given individual will be issued quarters. For an individual not occupying government quarters, a finance-office clerk then determines which rates for basic allowance for quarters ("singles" or the "with dependents") and geographically indexed variable housing allowance apply.

Compensation, other extrinsic rewards, and opportunities for intrinsically rewarding experiences are organizational resources. At each level of the process described above, discretion in distributing these resources is successively narrowed by a set of rules, policies, or procedures established at higher levels. New or expanded forms of differentiation identified in this study would give commanders and supervisors more discretion in distributing these resources. Before extending additional discretion, prudent legislators and high-level executive officials would ascertain that adequate accountability mechanisms exist to ensure satisfactory outcomes and avoid fraud, waste, and abuse. Specifically, accountability should be maintained to ensure that differentiated rewards are based on appropriate organizational outcomes, are free from intentional or unintentional distortions that benefit some individuals inappropriately, and meet public expectations for fairness, equal opportunity, and other similar standards.

ASSESSING ACCOUNTABILITY MECHANISMS

In the paragraphs that follow, we review accountability mechanisms available within DoD and discuss their adequacy for assessing differentiation in compensation.

Bureaucratic Mechanisms

The fundamental pattern of accountability at the operating level of government is bureaucratic. HRM innovations, however, can weaken this form of accountability. Empowerment of employees is, essentially, an expansion of discretion and a loosening of rules. Decentralization of functions is a distribution of decisionmaking authority and, typically, some measure of discretion to lower organizational levels. Additionally, organizational structures incorporating flatter hierarchies or nonhierarchical arrangements, or flexible workplace practices, tend to distance subordinates from the oversight of superiors. These structural changes weaken a pattern of accountability that depends on superiors directly monitoring their subordinates' work processes. Finally, HRM innovations that make rewards contingent on outputs or outcomes necessarily require superiors to rely on methods other than direct observation of process.

As organizations increasingly shed bureaucratic features that have served to ensure accountability, other patterns of accountability must be substituted (Romzek and Dubnick, 1994). Legal mechanisms and inspector general systems, for example, provide forms of control that can be independent of hierarchical relationships. Accounting and other information systems can provide details about processes, outputs, and outcomes that can be examined by stakeholders outside the organization, as well as by interested parties who are within the organization but outside of the responsible hierarchy. Political and public-interest mechanisms and professional standards provide additional mechanisms that are less dependent on bureaucratic or hierarchical relationships. In the following paragraphs, we discuss the suitability and applicability of each of these patterns within military organizations.

Legal Mechanisms

In the military case, legal mechanisms abound. Most federal statutes governing, for example, acquisitions, freedom of information, and protection of privacy apply to military as well as nonmilitary departments and agencies. Military personnel management is subject, along with civil service systems in general, to legislation and legal precedents governing equal employment opportunity and due process. Moreover, organizational behavior is governed extensively by military-specific legislation, such as that placing military organizations under civilian control or specifying many of the policies and structures governing grade and strength-management, promotion, separation, and retirement systems. Individual behavior is governed through a separate Uniform Code of Military Justice (UCMJ) specifically tailored to the conditions of military service.

The UCMJ, for example, criminalizes acts by military members (e.g., conduct unbecoming an officer) that, in a nonmilitary context, might be characterized merely as bad judgment or poor taste. Legal mechanisms permit interested parties, who may be either internal or external to military organizations, to use the authority of federal or military courts to sanction, restrain, or otherwise modify organizational or individual behavior. To the extent that bureaucratic forms of accountability might be diluted in military organizations, a shift to greater reliance on legal mechanisms is a possibility for enforcing accountability. It should be noted, however, that legal mechanisms, as with bureaucratic patterns, tend to focus on process rather than outputs or outcomes, to place constraints on HRM systems, and to inhibit innovation.

Inspectors General

Inspectors general and their staffs provide top-level commanders and managers with a broad oversight capability. They permit higher authorities to skip hierarchical levels in examining and evaluating issues in their organizations and also permit individuals to skip levels in seeking redress of grievances. Some inspectors general periodically and systematically inspect lower-level organizations. Others are limited, however, in that they depend on external stimuli to bring problems to their attention. Inspectors general functions can be useful in curbing fraud, waste, and abuse and in promoting practices that yield better outcomes.

Accounting Systems

Accounting systems should provide interested observers, who may be internal or external to an organization, a clear picture of assets, liabilities, revenues, costs, and other financial conditions of an organization. Unfortunately, in the opinion of the U.S. General Accounting Office (GAO) and other observers (Dodd, 1995), DoD financial management cannot provide such a picture. It is characterized by, among other things, contractor overpayments, improper payroll payments, disbursements not matched with related obligations, overexpended accounts, and an inability to determine actual operating costs for the Defense Business Operating Fund (Bowsher, 1994).

At lower organizational levels, utility of the system is equally impaired. In the experience of one of the authors of this report, efforts to devolve management of resources (e.g., temporary-duty travel funds or civilian-personnel hiring and promotion decisions) to lower-level unit commanders are hampered by financial management systems that are untimely, inaccurate, and lacking in sufficient detail.

Without significant improvements, the conditions for extending additional discretion over pay and other financial matters to lower organizational levels are problematic. Using standard accounting tools, managers would have difficulty basing differentiation of compensation on cost-related performance of their organizations because costs of specific activities cannot be reliably derived from DoD accounting systems. Moreover, they would have difficulty tracking their own disbursement of compensation resources. If they were to create their own local tools for these purposes, their management of differentiated-reward programs would be subject to very limited oversight and might not conform to accepted accounting practices.

Other Information Systems

Accounting systems are a special case of information systems, which can carry a variety of input, process, output, and outcome measures of organizational activity. The ability of networks to expand access to various functional databases (accounting, personnel, logistics, operations, etc.) increasingly makes such information available to more members of organizations at all levels. As information systems become reliable and complete, stakeholders inside and outside of organizations might find that access to databases, rather than formal reporting through hierarchies, provides an acceptable, if not superior, means of ensuring accountability.

However, at the present state of information-system technology in the DoD, this ideal is far from being realized. Services typically do not make their databases accessible to outsiders; functional communities within services typically do not make their databases accessible to other functional communities. In military personnel management, for example, DoD officials have access only to a small subset of service personnel databases, periodically extracted and provided to the Defense Manpower Data Center. Database "owners" can cite rational arguments for this restricted access to their data, including a fear that unsophisticated analysts will misinterpret data and use it to the detriment of the organization or function. Weber (1946, p. 16) warned, however, that "every bureaucracy seeks to increase the superiority of the professionally informed by keeping their knowledge and intentions secret." Bureaucracies protect information in order to promote their own interests by, for example, sheltering themselves from criticism and ensuring asymmetry of information in budgetary deliberations.

Political and Public-Interest Mechanisms

These mechanisms—including oversight by congressional committees and political appointees in the Executive Branch, vigilance by interest groups, lobbying, investigative reporting by news media, independent analysis by public-interest organizations, and whistle-blowing—provide additional avenues of accountability. Legislative oversight, in particular, provides an important means of extending broad discretion to organizational managers while maintaining public accountability. According to Schick (1987), "laws constrain administrative discretion in advance; oversight has the advantage of allowing full scope to administration, with legislative scrutiny occurring afterwards. The oversight role renders broad administrative discretion legitimate by making it accountable to legislative authority."

These mechanisms are generally functional at the higher levels of organization, or with respect to systemic issues. Their operation can be seen in such high-visibility issues as weapon system acquisition, fratricidal combat casualties, aircraft-safety investigations, sexual-harassment investigations, and flag officer promotions. However, they may have little potential to penetrate to the routine workings of military organizations and thus would not be expected to contribute significantly to accountability at lower organizational levels. The effectiveness of these mechanisms is a function of legislative staffs' and other outsiders' access to organizational data, which is generally limited.

Professional Standards

Public policy often embraces a deference to professional expertise, particularly in the traditional professions of law, medicine, and the clergy. Claims to professional status are also made by or on behalf of other occupational groups, including military officers (Thie and Brown et al., 1994; Mosher, 1987). Professions often claim, and are granted, authority to formulate and enforce their own standards and ethics on the grounds that only the members of a profession have the training and experience to evaluate the performance of their peers. Thus, adherence to professional standards provides an accountability mechanism that can be applied to military officers in their management of military affairs. Since administrative procedures are available to exclude from the profession (i.e., discharge from service) those who do not meet standards, this mechanism has teeth.

However, in matters less uniquely military, including compensation and other human resource management issues, military professionals have no special expertise and can claim little or no deference to their professional judgment. Thus, professional standards cannot function strongly as an accountability mechanism with respect to issues such as differentiation of compensation or other rewards.

DIFFICULTIES IN SHIFTING PATTERNS OF ACCOUNTABILITY IN MILITARY ORGANIZATIONS

Romzek and Dubnick (1994) argue that new flexibilities in public personnel management require a shift from bureaucratic accountability, which focuses on inputs, and legalistic mechanisms, which focus on process, toward political and professional mechanisms, which are more suitable for monitoring organizational outputs and outcomes. Without such a shift, they predict that new flexibilities are likely to be eroded by pressures to reassert bureaucratic and legalistic accountability mechanisms.

We see military organizations as not being well situated to shift patterns of accountability. As we noted earlier in this appendix, shifts to nonbureaucratic and nonhierarchical patterns rely, to a great extent, on open information systems. We do not see the necessary openness within the services. Where there is some openness, such as in accounting systems, we see inaccuracies and inadequacies sharply limiting the effectiveness of the oversight. We see professional military standards as not being specifically relevant to compensation and human resource management issues.

Shifting to alternative forms of accountability imposes some risk on subordinates. In a bureaucratic pattern of accountability, individuals are held accountable primarily for inputs over which they have extensive control—their own level of effort, quality of decisions, and compliance with rules. As shifts occur to alternative forms of accountability, outputs and outcomes rather than processes are observed. As Figure C.1 indicates, outputs depend on organizational inputs, inputs from other workers, and the functioning of a suitable process. Outcomes additionally depend on environmental factors. Individuals thus face the risk of being held accountable for results that are only partially within their control. Additionally, since assessment of outputs and outcomes can be more difficult than assessment of process, individuals risk being held accountable for inaccurately evaluated results. To the extent that compensation is made contingent on outputs or outcomes, organizations should expect to pay a risk premium in the form of higher expected compensation.

REFERENCES

Ackerlof, George A., "Gift Exchange and Efficiency-Wage Theory: Four Views," *American Economic Review*, Vol. 74, No. 2, May 1984, pp. 79–81.

Adams, J. Stacy, "Toward an Understanding of Inequity," *Journal of Abnormal and Social Psychology*, Vol. 67, No. 5, 1963, pp. 422–436.

Alchian, Armen, and H. Demsetz, "Production, Information Costs, and Economic Organization," *American Economic Review*, Vol. 62, December 1972, pp. 777–795.

Aldrich, Howard E., *Organizations and Environments*, Englewood Cliffs, NJ: Prentice Hall, 1979.

Anderson, Jon R., "Marine Grumbling About Tents Continues," *Air Force Times*, October 9, 1995.

Asch, B. J., and J. T. Warner, *A Theory of Military Compensation and Personnel Policy*, Santa Monica, CA: RAND, MR-439-OSD, 1994.

Atkinson, J. W., *An Introduction to Motivation*, New York: Van Nostrand, 1964.

Becker, Gary S., *Human Capital*, 2nd ed., New York: Columbia University Press, 1975.

Black, Matthew, *Personal Discount Rates: Estimates for the Military Population*, study prepared for the Fifth Quadrennial Review of Military Compensation, Arlington, VA: Systems Research and Applications Corporation, May 1983.

Blaker, James, "How the Pentagon Designs Its 21st Century Strategy," *The Christian Science Monitor*, January 30, 1996, p. 19.

Bowsher, Charles A., *Financial Control and System Weaknesses Continue to Waste DoD Resources and Undermine Operations*, testimony before the Committee on Governmental Affairs, U.S. Senate, Washington, D.C.: General Accounting Office, GAO/T-AIMD/NSIAD-94-154, April 12, 1994.

Braverman, Harry, *Labor and Monopoly Capital: The Degradation of Work in the Twentieth Century*, New York: Monthly Review Press, 1974.

Bridges, William P., and Wayne J. Villemez, *The Employment Relationship: Causes and Consequences of Modern Personnel Administration*, New York: Plenum Press, 1994.

Builder, Carl H., *The Masks of War: American Military Styles in Strategy and Analysis*, Baltimore, MD: Johns Hopkins University Press, 1989.

Callander, Bruce D., "The New Way of Officer Assignments," *Air Force Magazine*, September 1995.

Campbell, J. P., M. D. Dunnette, Edward E. Lawler, and K. E. Weick, *Managerial Behavior, Performance, and Effectiveness*, New York: McGraw-Hill, 1970.

Cascio, Wayne F., "Whither Industrial and Organizational Psychology in a Changing World of Work?" *American Psychologist*, Vol. 50, No. 11, November 1995, pp. 928–939.

Clinton, Bill, *Memorandum for the Secretary of Defense, Subject: The Eighth Quadrennial Review of Military Compensation*, Washington, D.C.: The White House, January 27, 1995.

Cooper, Richard V. L., *The All-Volunteer Force and Defense Manpower—Testimony Before the House Budget Committee (Task Force on National Security), July 12, 1977*, Santa Monica, CA: RAND, P-5911, 1977.

Cowherd, Douglas M., and David I. Levine, "Product Quality and Pay Equity Between Lower-Level Employees and Top Management: An Investigation of Distributive Justice Theory," *Administrative Science Quarterly*, Vol. 37, No. 2, June 1992, pp. 302–320.

Crocker, Keith J., and Scott E. Masten, "Mitigating Contractual Hazards: Unilateral Options and Contract Length," *RAND Journal of Economics*, Vol. 19, 1988, pp. 327–343.

———, "Pretia Ex Machina? Prices and Process in Long-Term Contracts," *Journal of Law and Economics*, Vol. 34, 1991, pp. 66–99.

———, "Regulation and Administered Contracts Revisited: Lessons from Transaction-Cost Economics for Public Utility Regulation," *Journal of Regulation Economics*, Vol. 9, January 1996, pp. 5–39.

Crocker, Keith J., and Kenneth J. Reynolds, "The Efficiency of Incomplete Contracts: An Empirical Analysis of Air Force Engine Procurement," *RAND Journal of Economics*, Vol. 24, Spring 1993, pp. 126–146.

Crosby, Faye, Laura Burris, Catherine Censor, and E. R. MacKethan, "Two Rotten Apples Spoil the Justice Barrel," in Hans Werner Bierhoff, Ronald L. Cohen, and Jerald Greenberg, eds., *Justice in Social Relations*, New York: Plenum, 1986.

Dahlman, Carl, "The Problem of Externality," *Journal of Law and Economics*, Vol. 22, 1979, pp. 141–162.

Delzompo, F. A., "The Few, the Proud, the Unwed," *Naval Institute Proceedings*, November 1995, pp. 48–49.

Dittrich, John E., and Michael R. Carrell, "Organizational Equity Perceptions, Employee Job Satisfaction, and Departmental Absence and Turnover Rates," *Organizational Behavior and Human Performance*, Vol. 24, 1979, pp. 29–40.

Dodd, Joseph G., *The Chief Financial Officers Act of 1990: Conserving Resources for Readiness and National Security Through Better Financial Management*, Washington, D.C.: Industrial College of the Armed Forces, Executive Research Project S7, 1995.

Doeringer, Peter B., and Michael J. Piore, *Internal Labor Markets and Manpower Analysis*, Lexington, MA: D. C. Heath, 1971.

Dresang, Dessis L., *Public Personnel Management and Public Policy*, New York: Longman, 1991.

Edwards, Richard, *Contested Terrain: The Transformation of the Workplace in the Twentieth Century*, New York: Basic Books, 1979.

Fayol, Henri, *General and Industrial Management*, London: Pitman, 1949 trans. (first published in 1919).

Festinger, Leon, *A Theory of Cognitive Dissonance*, Stanford, CA: Stanford University Press, 1957.

Goetz, C. J., and R. E. Scott, "Principles of Relational Contracts," *Virginia Law Review*, Vol. 67, 1981, pp. 1089–1150.

Greenberg, Jerald, "A Taxonomy of Organizational Justice Theories," *Academy of Management Review*, Vol. 12, No. 1, 1987, pp. 9–22.

Hamel, G., and Prahalad, C. K., *Competing for the Future*, Boston, MA: Harvard Business School Press, 1994.

Hannan, Michael T., and John Freeman, "The Population Ecology of Organizations," *American Journal of Sociology*, Vol. 82, March 1977, pp. 929–964.

Hatry, Harry P., and Donald M. Fisk, "Measuring Productivity in the Public Sector," in Marc Holzer, ed., *Public Productivity Handbook*, New York: Marcel Dekker, 1992, pp. 139–160.

Herzberg, Frederick, "One More Time: How Do You Motivate Employees?" *Harvard Business Review*, Vol. 46, No. 1, 1968.

Holmstrom, Bengt, "Moral Hazard and Observability," *Bell Journal of Economics*, Vol. 10, No. 1, Spring 1979, pp. 74–91.

Hosek, James R., Christine E. Peterson, Jeannette Van Winkle, and Hui Wang, *A Civilian Wage Index for Defense Manpower*, Santa Monica, CA: RAND, R-4190-FMP, 1992.

Ivancevich, J. M., A. D. Szilagyi, and M. J. Wallace, *Organizational Behavior and Performance*, Belmont, CA: Goodyear, 1977.

Jordan, Bryant, "Promotion Folder Change: Advanced Degrees Lose Luster for Captains, Majors," *Air Force Times*, January 15, 1996.

Joskow, P. L., "Contract Duration and Relationship-Specific Investments: Empirical Evidence from Coal Markets," *American Economic Review*, Vol. 77, 1987, pp. 168–185.

Katzenbach, J. R., and Smith, D. K., *The Wisdom of Teams: Creating the High-Performance Organizations*, Boston, MA: Harvard Business School Press, 1993.

Keppel, Geoffrey, and Sheldon Zedeck, *Data Analysis for Research Designs*, New York: W. H. Freeman, 1989.

Kerr, Clark, "The Balkanization of Labor Markets," in Paul Webbink, ed., *Labor Mobility and Economic Opportunity*, New York: John Wiley & Sons, 1954, pp. 92–110.

Lawrence, Paul R., and Jay W. Lorsch, *Organization and Environment: Managing Differentiation and Integration*, Boston: Harvard University, 1967.

Lawler, Edward E., *Strategic Pay: Aligning Organizational Strategies and Pay Systems*, San Francisco: Jossey-Bass, 1990.

Lazear, Edward P., *Personnel Economics*, Cambridge, MA: MIT Press, 1995.

Lazear, Edward P., and Sherwin Rosen, "Rank-Order Tournaments as Optimum Labor Contracts," *Journal of Political Economy*, Vol. 89, No. 5, 1981, pp. 841–864.

Lind, E. Allan, and Tom R. Tyler, *The Social Psychology of Procedural Justice*, New York: Plenum Press, 1988.

March, James G., and Herbert A. Simon, *Organizations*, New York: John Wiley, 1958.

Martin, Joanne, and Alan Murray, "Distributive Injustice and Unfair Exchange," in David M. Messick and Karen S. Cook, eds., *Equity Theory: Psychological and Sociological Perspectives*, New York: Praeger, 1983, pp. 169–206.

Maser, Stephen M., "Transaction Costs in Public Administration," in Donald J. Calista, ed., *Public Policy Studies: Volume 9, Bureaucratic and Government Reform*, Greenwich, CT: JAI Press, 1986, pp. 55–71.

Maslow, Abraham H., "A Theory of Human Motivation," *Psychological Review*, Vol. 50, 1943, pp. 370–396.

Masten, Scott E., "Transaction Costs, Mistakes, and Performance: Assessing the Importance of Governance," *Managerial and Decision Economics*, Vol. 14, 1993, pp. 119–129.

Mayer, Kenneth R., and Anne M. Khademian, "Bringing Politics Back In: Defense Policy and the Theoretical Study of Institutions and Processes," *Public Administration Review*, Vol. 56, No. 2, 1996, pp. 180–190.

Mayo, Elton, *The Social Problems of an Industrial Civilization*, Boston: Harvard University, 1945.

McClelland, D. C., *The Achieving Society*, Princeton, NJ: Van Nostrand, 1961.

McGregor, Douglas, *The Human Side of Enterprise*, New York: McGraw-Hill, 1960.

Monteverde, Kirk, and David Teece, "Supplies, Switching Costs, and Vertical Integration in the Automobile Industry," *Bell Journal of Economics*, Vol. 13, Spring 1982, pp. 206–213.

Moore, S. Craig, et al., *Measuring Military Readiness and Sustainability*, Santa Monica, CA: RAND, R-3842-DAG, 1991.

Mortenson, Dale T., "Specific Capital and Labor Turnover," *RAND Journal of Economics*, Vol. 9, Autumn 1978, pp. 572–586.

Mosher, Frederick C., "The Professional State," in Dean L. Yarwood, ed., *Public Administration: Politics and the People*, White Plains, NY: Longman, 1987.

Muris, T. J., "Opportunistic Behavior and the Law of Contracts," *Minnesota Law Review*, Vol. 65, 1981, pp. 521–590.

Nichiporuk, Brian, and Carl H. Builder, *Information Technologies and the Future of Land Warfare*, Santa Monica, CA: RAND, MR-560-A, 1995.

Oldham, Greg R., Carol T. Kulik, Maureen L. Ambrose, Lee P. Stepina, and Julianne F. Brand, "Relations Between Job Facet Comparisons and Employee Reactions," *Organizational Behavior and Human Decision Processes*, Vol. 38, 1986, pp. 28–47.

Orvis, Bruce R., James R. Hosek, and Michael G. Mattock, with Rebecca M. Mazel and Iva S. MacLennan, *PACER SHARE Productivity and Personnel Management Demonstration: Third-Year Evaluation*, Santa Monica, CA: RAND, MR-310-P&R, 1993.

Ouchi, William G., "Markets, Bureaucracies, and Clans," *Administrative Science Quarterly*, Vol. 25, 1980, pp. 120–142.

Pfeffer, Jeffrey, and Nancy Langton, "The Effect of Wage Dispersion on Satisfaction, Productivity, and Working Collaboratively: Evidence from College and University Faculty," *Administrative Science Quarterly*, Vol. 38, No. 3, 1993, pp. 382–407.

Pfeffer, Jeffrey, and Gerald R. Salancik, *The External Control of Organizations*, New York: Harper & Row, 1978.

Porter, Lyman W., and Edward E. Lawler III, *Managerial Attitudes and Performance*, Homewood, IL: Richard D. Irwin, Inc., and The Dorsey Press, 1968.

Porter, Lyman W., Edward E. Lawler III, and J. Richard Hackman, *Behavior in Organizations*, New York: McGraw-Hill, 1975.

Pressman, Jeffrey L., and Aaron Wildavsky, *Implementation*, Berkeley: University of California Press, 1973.

Pritchard, Robert D., Marvin D. Dunnette, and Dale O. Jorgenson, "Effects of Perceptions of Equity and Inequity on Worker Performance and Satisfaction," *Journal of Applied Psychology*, Vol. 56, 1972, pp. 75–94.

Romzek, Barbara S., and Melvin J. Dubnick, "Issues of Accountability in Flexible Personnel Systems," in Patricia W. Ingraham and Barbara S. Romzek, eds., *New Paradigms for Government*, San Francisco: Jossey-Bass, 1994.

Rosen, Sherwin, "Prizes and Incentives in Elimination Tournaments," *American Economic Review*, Vol. 76, No. 4, 1986, pp. 701–715.

———, "The Theory of Equalizing Differences," in Orley Ashenfelter and Richard Layard, eds., *Handbook of Labor Economics*, Vol. I, New York: Elsevier Science Publishers, 1986, pp. 641–692.

Ross, Stephen A., "The Economic Theory of Agency: The Principal's Problem," *American Economic Review*, Vol. 43, No. 2, 1973, pp. 134–139.

Rumsey, Michael G., "The Best They Can Be: Tomorrow's Soldiers," U.S. Army Research Institute paper and briefing delivered at the Army 2010 Conference (Future Soldiers and the Quality Imperative), Cantigny Conference Center, Wheaton, IL, May 31–June 2, 1995.

Sashkin, Marshall, and Richard L. Williams, "Does Fairness Make a Difference?" *Organizational Dynamics*, Vol. 19, No. 2, 1990, pp. 56–71.

Schick, Allen, "Congress and the 'Details' of Administration," in Dean L. Yarwood, ed., *Public Administration: Politics and the People*, White Plains, NY: Longman, 1987.

Scott, Richard W., *Organizations: Rational, Natural, and Open Systems*, 2nd ed., Englewood Cliffs, NJ: Simon & Schuster, 1987.

Selznick, Philip, "Foundations of the Theory of Organization," *American Sociological Review*, Vol. 13, February 1948, pp. 25–35.

Seventh Quadrennial Review of Military Compensation, *Report of the Seventh Quadrennial Review of Military Compensation*, Washington, D.C.: OASD(FMP), 1992.

Sheppard, Blair H., Roy J. Lewicki, and John W. Minton, *Organizational Justice: The Search for Fairness in the Workplace*, New York: Lexington Books, 1992.

Simon, Herbert A., *Administrative Behavior*, 3rd ed., New York: Free Press, 1976 (1st ed., 1945).

Skinner, B. F., *The Behavior of Organisms: An Experimental Analysis*, New York: Appleton-Century-Crofts, 1938.

Steers, Richard M., and Richard T. Mowday, "Employee Turnover and Post-Decision Accommodation Processes," in L. L. Cummings and Barry M. Staw, eds., *Research in Organizational Behavior*, Vol. 3, Greenwich, CT: JAI Press, 1981, pp. 235–281.

Stiglitz, Joseph E., "The Efficiency Wage Hypothesis, Surplus Labour, and the Distribution of Income in L.D.C.s," *Oxford Economic Papers*, Vol. 28, No. 2, July 1976, pp. 185–207.

Summers, Timothy P., and William H. Hendrix, "Modelling the Role of Pay Equity Perceptions: A Field Study," *Journal of Occupational Psychology*, Vol. 64, 1991, pp. 145–157.

Taubman, Paul, and Michael L. Wachter, "Segmented Labor Markets," in Orley Ashenfelter and Richard Layard, eds., *Handbook of Labor Economics*, Vol. II, New York: Elsevier Science Publishers, 1986, pp. 1183–1217.

Taylor, Frederick W., *The Principles of Scientific Management*, New York: Harper, 1911.

Telly, Charles S., Wendell L. French, and William G. Scott, "The Relationship of Inequity to Turnover Among Hourly Workers," *Administrative Science Quarterly*, Vol. 16, 1971, pp. 164–172.

Thie, Harry J., Roger A. Brown, et al., *Future Career Management Systems for U.S. Military Officers*, Santa Monica, CA: RAND, MR-470-OSD, 1994.

Toffler, Alvin, and Heide Toffler, *War and Anti-War: Survival at the Dawn of the Twenty-First Century*, Boston: Little, Brown, 1993.

———, *Creating a New Civilization: The Politics of the Third Wave*, Atlanta: Turner Publishing, 1994.

Trice, Harrison M., and Janice M. Beyer, *The Cultures of Work Organizations*, Englewood Cliffs, NJ: Prentice Hall, 1993.

Uniformed Services Almanac, Falls Church, VA: Uniformed Services Almanac, Inc., 1995.

Vroom, Victor H., *Work and Motivation*, New York: John Wiley & Sons, 1964.

Wageman, Ruth, "Interdependence and Group Effectiveness," *Administrative Science Quarterly*, Vol. 40, 1995, pp. 145–180.

Watson, John B., *Psychology from the Standpoint of a Behaviorist*, Philadelphia, PA: Lippincott, 1919.

Weber, Max, *From Max Weber: Essays in Sociology*, Hans H. Gerth and C. Wright Mills, eds. and trans., New York: Oxford University Press, 1946 (first published in 1906–1924).

White, John, et al., *Directions for Defense: Report of the Commission on Roles and Missions of the Armed Forces*, Washington, D.C.: Department of Defense, May 24, 1995.

Williamson, Oliver E., *Markets and Hierarchies: Analysis and Antitrust Implications*, New York: Free Press, 1975.

———, *The Economic Institutions of Capitalism: Firms, Markets, Relational Contracting*, New York: Free Press, 1985.

———, *Handbook of Industrial Organization*, New York: Elsevier Science Inc., 1989, pp. 136–182.

———, "Comparative Economic Organizational Analysis of Discrete Market Alternatives," *Administrative Science Quarterly*, Vol. 36, 1991, pp. 269–296.

Yellen, Janet L., "Efficiency Wage Models of Unemployment," *American Economic Review*, Vol. 74, No. 2, May 1984, pp. 200–205.